Cambridge Planetary Science Series

Editors: W. I. Axford, R. Greeley, G. Hunt

Remote sounding of atmospheres

T0296324

Remote sounding of atmospheres

J. T. HOUGHTON

Director General, Meteorological Office, Bracknell

F. W. TAYLOR

Head, Department of Atmospheric Physics, University of Oxford

C. D. RODGERS

Lecturer, Department of Atmospheric Physics, University of Oxford

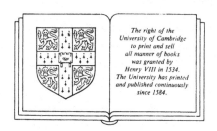

The right of the
University of Cambridge
to print and sell
all manner of books
was granted by
Henry VIII in 1534.
The University has printed
and published continuously
since 1584.

CAMBRIDGE UNIVERSITY PRESS

Cambridge

London   New York   New Rochelle

Melbourne   Sydney

CAMBRIDGE UNIVERSITY PRESS
Cambridge, New York, Melbourne, Madrid, Cape Town, Singapore, São Paulo, Delhi

Cambridge University Press
The Edinburgh Building, Cambridge CB2 8RU, UK

Published in the United States of America by Cambridge University Press, New York

www.cambridge.org
Information on this title: www.cambridge.org/9780521310659

First published 1984
First paperback edition 1986
Re-issued in this digitally printed version 2009

A catalogue record for this publication is available from the British Library

Library of Congress Cataloguing in Publication data
Houghton, John Theodore.
Remote sounding of atmospheres.

(Cambridge planetary science series)
Bibliography: p.
Includes index.
1. Atmosphere – Remote sensing. I. Taylor, F. W.
II. Rodgers, C. D. III. Title. IV. Series.
[QC871.H78 1986]   551.5′1′0287   85-29084

ISBN 978-0-521-24281-3 hardback
ISBN 978-0-521-31065-9 paperback

CONTENTS

PREFACE

Remote sounding has grown in importance very rapidly in recent years to become one of the most important experimental techniques in the earth and planetary sciences. This can be attributed to two main reasons: the much wider availability of space platforms from which to make the observations, and the tremendous growth in computer power available to the scientist who needs to manipulate and analyse the very large amounts of data involved. The recent rapid progress in both areas shows no signs of abating in the years ahead and the subject seems bound to increase in popularity even more.

As in most growth areas there is a need for basic literature on the subject to cater for the expanding numbers of workers in the field. The present text is based on a review paper by two of us which appeared in the journal *Reports on Progress in Physics* in 1973. There, we presented the basic principles of remote sounding and described the applications which had been made to the study of the atmospheres of the Earth and planets. This new volume not only brings these aspects up to date, but also adds sections on the fundamentals of retrieval theory and on the assimilation of remote sounding data into meteorological problems. These additions and improvements are intended to produce a comprehensive text-book for the student or researcher entering the field of atmospheric remote sounding for the first time.

We would like to express our appreciation to our many colleagues who have contributed to the developments which we describe in the following pages. We are particularly grateful to those authors and publishers who gave permission for their diagrams and illustrations to be reproduced. Our special thanks go to Miss C. M. Wagstaff for typing the manuscript.

> J. T. Houghton
> F. W. Taylor
> C. D. Rodgers
> *Oxford, March 1983*

1
INTRODUCTION

1.1 Definitions

Remote sounding techniques involve the measurement of atmospheric parameters from a distance. This definition includes the study of our own atmosphere from above, by the use of instruments on artificial satellites, the exploration of other planets by 'flyby' or orbiting probes, and ground-based observations of planets or the higher parts of the Earth's atmosphere using telescopes where necessary. It also includes the use of rockets, balloons and high-flying aircraft as platforms for the same kind of observation. The feature which all of these have in common is that the instrument making the measurement is far from the point at which the measurement is being made. This is in contrast with, for example, routine meteorological techniques for the study of the upper air, performed by injecting instruments into the interesting region using balloons or rockets and in which no measurements are made in parts of the atmosphere which the payload does not pass through.

Another example comes from the planetary field; two different approaches have been employed in the exploration of Venus. On the one hand remote sounding techniques have been used from spacecraft which fly past or orbit the planet, and on the other hand direct measurements have been made from probes which enter the atmosphere and fall to the surface.

Remote sounding is not a new concept. Until very recently, all of our knowledge of the atmospheres of other planets came from observations made from the Earth, mostly with instruments coupled to large telescopes on the ground. Much early research on the Earth's upper atmosphere was pursued by remote techniques. Observations of the anomalous propagation of sound suggested the presence of a higher temperature region in the stratosphere; this hypothesis was confirmed by Lindemann and Dobson's observations of meteor trails which led to Dobson's classic investigation of ozone. Further, the ionosphere was discovered through studies of radio propagation. All of these early investigations of the upper atmosphere during the period 1920-40 employed remote techniques. Since 1940 the bulk of our additional knowledge, particularly about the neutral upper atmosphere, has come from more direct sounding from balloons and rockets. The new feature which accounts for the currently enhanced interest in remote techniques is the availability of much better platforms from which to exploit them, namely, of course, satellites and interplanetary spacecraft. For the earth, it is much easier to study the properties of all but the lowest layers of the atmosphere from above, looking down, than it is from below looking up. In addition, the vastly increased coverage offered by a satellite makes it possible to look at complete weather systems for the first time, and to follow their progress and evolution around the globe. For planetary exploration, spacecraft make remote studies much more productive for two main reasons. Firstly, the intervening blanket of the Earth's atmosphere is no longer present. Secondly, moving close to the planet makes for much better spatial resolution and allows complete phase coverage. Planetary observations from balloons, rockets, aircraft or Earth satellites (natural or artificial, manned or unmanned) offer the first advantage but not the second.

There is great scope for ingenuity in the design of instrumentation and the interpretation of measurements. Use has already been made of spectrometers, interferometers, radiometers, photometers and cameras, both of conventional and radical design, operating over a wide range of wavelengths ranging from the ultraviolet through the visible and infrared to the microwave and radio regions of the spectrum. In this book, confining ourselves to remote measurements from artificial satellites and space probes, we shall look at the physical basis for the various techniques, describing some methods which have been successfully applied, and at some of the more significant instruments and the results which they have obtained.

1.2   Objectives of measurements

At the present time, knowledge of planetary atmospheres other than our own is scant and most remote sounding investigations are directed towards the acquisition of fundamental information on composition, gross circulation, temperature-pressure structure, and the nature and behaviour of clouds. Radiative transfer and thermal balance are also important: this involves a study of how the incoming solar and outgoing thermal radiation fields interact with the atmospheric system in order to produce the observed structure and motions. In the case of the Jovian planets it is found that the total radiation leaving for space is considerably in excess of that arriving from the sun, which makes radiation balance a particularly interesting problem and also points to a need for experiments designed to reveal the source of the extra energy and its effect on the general circulation. On Venus, the processes which maintain the day to night temperature contrasts in the thermosphere, the polar stratospheric warming and the rapid zonal motions at the cloud tops are very poorly understood. Many other examples could be cited where fundamental information is needed.

Second-order objectives centre around a more detailed understanding of the circulation of the other planets, of the forces which drive them, and of interesting phenomena like precipitation, storms and lightning, atmospheric photochemistry, the polar vortex on Venus and the Great Red Spot on Jupiter. A good understanding of the variety of individual atmospheric types will lead eventually to an integrated, well-founded faculty in atmospheric science, atmospheric dynamics and planetary meteorology.

Exploratory measurements of our own atmosphere are by no means a thing of the past although we naturally have a more complex knowledge of this than we do of those of other worlds.

Fig. 1.1 illustrates the broad structure of the Earth's atmosphere. Below 10–15 km in altitude is the troposphere (turning-region), the seat of surface weather, where the temperature falls with altitude. A peak in temperature, the stratopause, is found at about 50 km and arises because of the absorption of solar radiation by ozone. Below this peak down to about 15 km is the stratosphere (layered region) and above it up to the mesopause at about 85 km is the mesosphere (middle-sphere). Above the mesosphere the temperature rises rapidly with altitude in the thermosphere. Reference will be made to these different atmospheric regions in later chapters.

What is required of atmospheric measurements is that as complete a description as possible be obtained of the fields of atmospheric density, motion and composition over the whole globe and at all heights of importance (Fig. 1.2). The net input of energy at the top of the atmosphere also needs to be measured as does the exchange of heat, momentum and water vapour with the surface.

Because of the advent on the one hand of satellites to provide the coverage and on the other hand of very large computers on which numerical models of atmospheric processes can be run, atmospheric scientists throughout the world are cooperating in a very large international enterprise, the Global Atmospheric Research Programme (GARP). The First GARP Global Experiment (FGGE), also known as the Global Weather Experiment, which was carried out in 1979 had the following aims, illustrated in Fig. 1.3:

(1) to provide a set of global measurements sufficiently accurate to provide initial conditions for a numerical model;

(2) to develop sophisticated numerical models of atmospheric motion;

(3) to test the validity of the models by comparing their predictions with the real atmosphere.

If our understanding of the organization of atmospheric motions can develop sufficiently so that they can be correctly represented in the models, and if sufficiently accurate data can be collected, it is the belief of meteorologists that prediction of the change and development of weather systems will be possible up to perhaps two weeks ahead and that eventually we may begin to understand some of the mechanisms which underlie climatic change.

Measurements are required to provide the initial data (Fig. 1.3) and also to provide diagnostic information to test the model and its various approximations. Table 1.1 summarizes the information required for the GARP (World

Fig. 1.1. Temperature (K) cross-section of the atmosphere from 80°N to 80°S as deduced from radiance measurements from the selective chopper radiometer on Nimbus 5 and the pressure modulator radiometer on Nimbus 6 for 4 August 1975. See §§ 6.7 and 6.8 for descriptions of the radiometers.

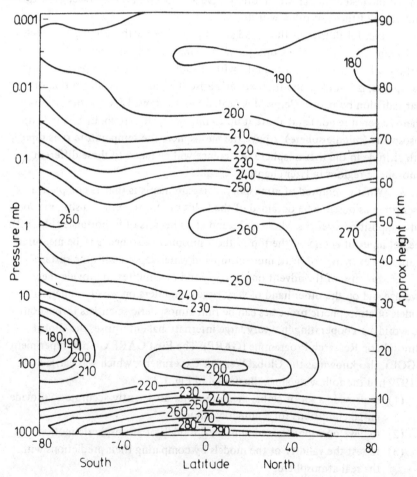

Table 1.1 *Observational requirements for the GARP.*

Atmospheric state parameters	Accuracy (RMS error)
Wind components	$\pm 3 \text{ m s}^{-1}$
Temperature	$\pm 1\,^{\circ}\text{C}$
Pressure of reference level	$\pm 0.3\%$
Water vapour pressure	± 1 mbar
Sea surface temperature	$\pm 0.25\,^{\circ}\text{C}$
Space and time averages	
Time average interval	2 hours
Horizontal space average	100 km
Vertical space average	Defined by the requirement of a minimum of 8 data levels at surface, 900, 700, 500, 200, 100, 50, 10 mbar respectively

Further information for verification purposes
Precipitation
Cloud-cover data
Snow or ice cover
Elements of radiation budget

Fig. 1.2. The principal parameters we need to know.

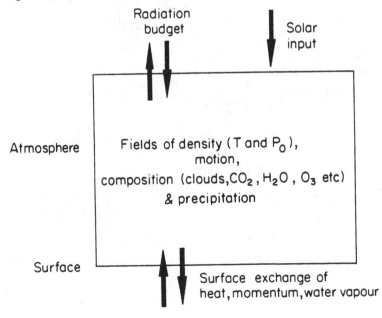

Fig. 1.3 The Global Weather Experiment.

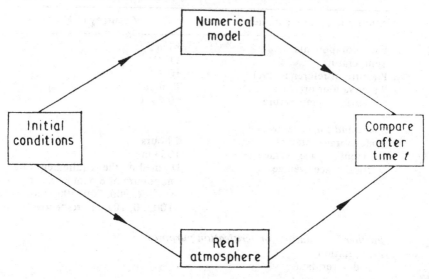

Fig. 1.4 Schematic illustration of the interaction of radiation with atmosphere.

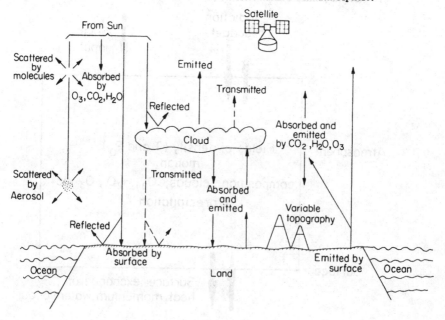

Meteorological Organization 1969). It is only by the use of remote sounding instruments on satellites that such a vast amount of data can be collected in the necessary time and at reasonable cost.

1.3   The general problem

All remote sounding techniques involve the detection and measurement of electromagnetic energy after it has been emitted from, reflected from, partially absorbed by or otherwise modified by the part of the atmosphere of interest. The problem of interpretation of such measurements is made difficult by the fact that the flux intensity arriving at the instrument is, in general, a function of many unknowns, and made worse by the fact that the dependence is often complex and may be only partly understood. Fig. 1.4 shows schematically how the flux from the sun can arrive at an instrument on a satellite after following many different routes. Before it reaches the detector, the radiation may have been scattered, reflected, absorbed and re-emitted several times, each time in a way that is characteristic of the composition, temperature and pressure of the atmospheric gases and the physical properties of the surface, clouds and aerosols. Furthermore, radiation from all parts of the atmosphere which are in the field of view will enter the instrument, and the question of obtaining vertical resolution (i.e. distinguishing between different height levels in the atmosphere) becomes a problem. For a planetary atmosphere in which the gross temperature–pressure structure and the composition is hardly known, especially if scattering and cloud effects are important, it becomes impossible to draw definite conclusions from a small number of measurements. The best that can be done is to reject theories and models which are incompatible with measurements which are known to be good. As more and better measurements are accumulated, a single acceptable model should begin to emerge. The planning and interpretation of subsequent experiments then become easier.

Much attention has been given in the last ten years to the theory of inversion or interpretation of measurements, particularly those made in the Earth's lower atmosphere for meteorological purposes. These often call for greater accuracy than would be acceptable in the sounding of planets or the less well understood parts of our own atmosphere. To achieve the stated objective of the GARP, for example (Table 1.1), good spatial resolution in three dimensions and a high radiometric accuracy are needed at the same time. Then the theory must be able to account fairly precisely for the origin of the flux which is measured and its dependence on the atmospheric parameters. This calls for highly developed theories of radiative transfer, molecular absorption and matter–radiation interaction and large amounts of laboratory data on the physical and spectral properties of the important gases, liquid droplets and solid particles, much of which is still not available.

Notwithstanding these difficulties, since the advent of satellites and probes the careful design of experiments to minimize problems and the use of models and approximations, wherever exact methods could not be used, has led to the acquisition of a remarkable amount of new information about our own and other atmospheric systems.

In the chapters which follow, first a description will be given of the various spacecraft which have been employed for remote sounding observations of the atmospheres of the earth and the planets. A number of chapters will then be devoted to measurements on the Earth's atmosphere including its radiation budget, temperature and composition. The theory of methods of retrieval of atmospheric parameters from remote radiance observations – sometimes called the inverse problem – is also presented. The final chapters are a description of the exploration of atmospheres of planets other than Earth by remote sensing from deep-space probes.

2

PLATFORMS IN SPACE

2.1 Polar-orbiting satellites

To provide near global coverage of the atmosphere a satellite in a near polar orbit can be employed. If the orbital altitude is constant at 1000 km then the satellite passes over each pole about fifteen times each day, the plane of the orbit remaining fixed in space while the earth rotates beneath it. If the inclination of the orbit is slightly different from 90°, then, because the earth is not a perfect uniform sphere, the plane of the orbit will precess in space. If the precession is arranged to be 360° per year, the phase of the orbit will always remain the same relative to the sun. This is called a sun-synchronous orbit. It is useful in atmospheric science because, whichever quantity is sounded, it is measured at each location with the sun always in the same place in the sky. Thus one important variable is fixed and changes which are observed to take place must be independent of this. Nimbus satellites have been flown in noon/midnight sun-synchronous orbits for which the satellite always crosses the equator at local noon or local midnight. The pattern traced over the earth by such a satellite is shown in Fig. 2.1. Several identical satellites deployed in different sun-synchronous orbits enable continuous coverage of the whole planet to be achieved with observations at each point repeated every few hours.

The first series of satellites designed specifically for atmospheric studies was TIROS (for Television and Infrared Observations Satellite), a cylindrical eighteen-sided polyhedron about 60 cm high and 1 m across, weighing about 170 kg (Fig. 2.2). TIROS was spin-stabilized so that its axis was fixed in space rather than relative to the earth's surface, a fairly serious limitation which resulted from engineering constraints in those early days. TIROS wheel satellites were improved versions which have the spin axis perpendicular to the plane of the orbit, so that instruments mounted on the rim view the earth perpendicularly once for each rotation of the satellite instead of once each orbit as previously. The later TIROS series satellites, designated ESSA (Environmental Survey Satellite), are wheel-type spacecraft in sun-synchronous orbits.

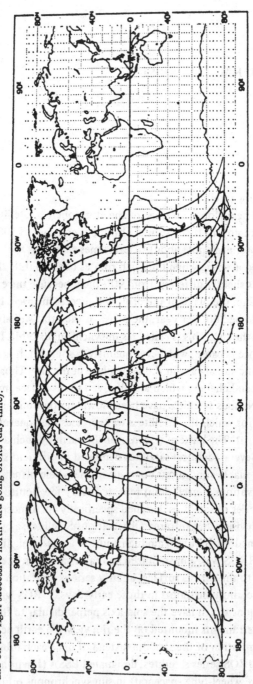

Fig. 2.1 Tracks of subsatellite point for polar-orbiting satellite Nimbus 4. On the left are shown successive southward going orbits (night-time) and on the right successive northward going orbits (day-time).

Nimbus is a series of experimental meteorological satellites (Nimbus being the Latin word for cloud). Experimental satellites are used as testbeds for new techniques, the most useful of which become standard equipment on operational satellites such as ESSA or ITOS (see below). The Nimbus spacecraft (Fig. 2.3) is larger (about 3.5 m high), heavier (500 kg), provides more power (450 W instead of 30 W) and is more advanced than TIROS. In particular it is earth-stabilized so that its axis is always perpendicular to the surface, using horizon sensors to keep it this way. All Nimbus satellites occupy polar sun-synchronous orbits at altitudes around 1000 km.

Following experience on the early Nimbus satellites, an improved TIROS operational system (ITOS) was developed, the first of the new series being launched early in 1970. Three-axis stabilization was employed similar to the Nimbus satellites in sun-synchronous orbits at 1460 km altitude.

Five ITOS satellites were launched by the National Oceanographic and Atmospheric Administration (NOAA) of the US. NOAA 2, for example, carried three main instruments, two of them scanning radiometers, each with two channels, one in the visible and one in the infrared window (see Ch.3) and a

Fig. 2.2 TIROS satellite. NASA illustration.

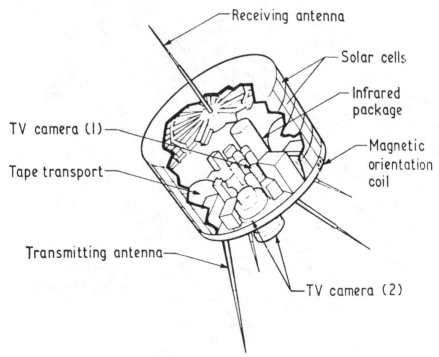

vertical temperature profile radiometer VTPR (see Ch.6). Its construction consists of a nearly cubical box, about 1 m wide and weighing about 300 kg,
attached to which are a pair of solar paddles providing about 150 W mean power
(see Fig. 2.4).

The launch of TIROS-N in 1978 by the US introduced the third generation of operational, polar-orbiting satellites. For this satellite, illustrated in Fig.
2.5, France provided the data collection system and the UK provided the stratospheric sounding unit. TIROS-N includes a microwave sounding unit for the first

Fig. 2.3 Nimbus 4 satellite. From NASA 4 User's Guide.

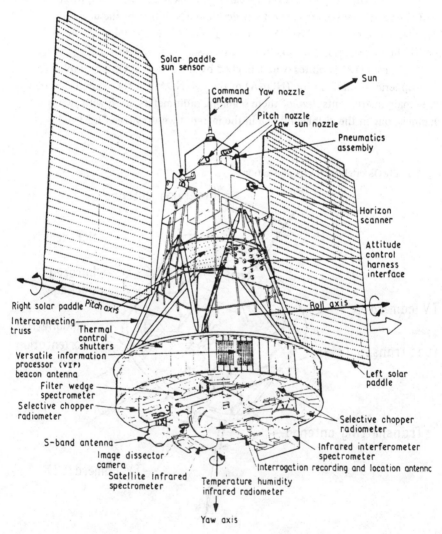

Fig. 2.4 ITOS satellite D. From Albert (1968).

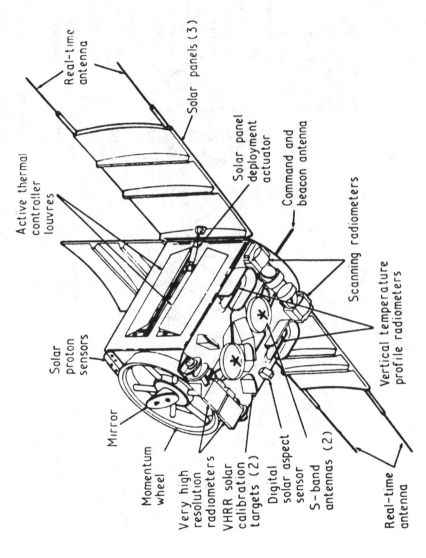

Real-time antenna

Solar panels (3)

Active thermal controller louvres

Solar panel deployment actuator

Command and beacon antenna

Scanning radiometers

Solar proton sensors

Vertical temperature profile radiometers

Mirror

Momentum wheel

Very high resolution radiometers

VHRR solar calibration targets (2)

Digital solar aspect sensor

S-band antennas (2)

Real-time antenna

Fig. 2.5 TIROS-N spacecraft. From Schwalb (1978).

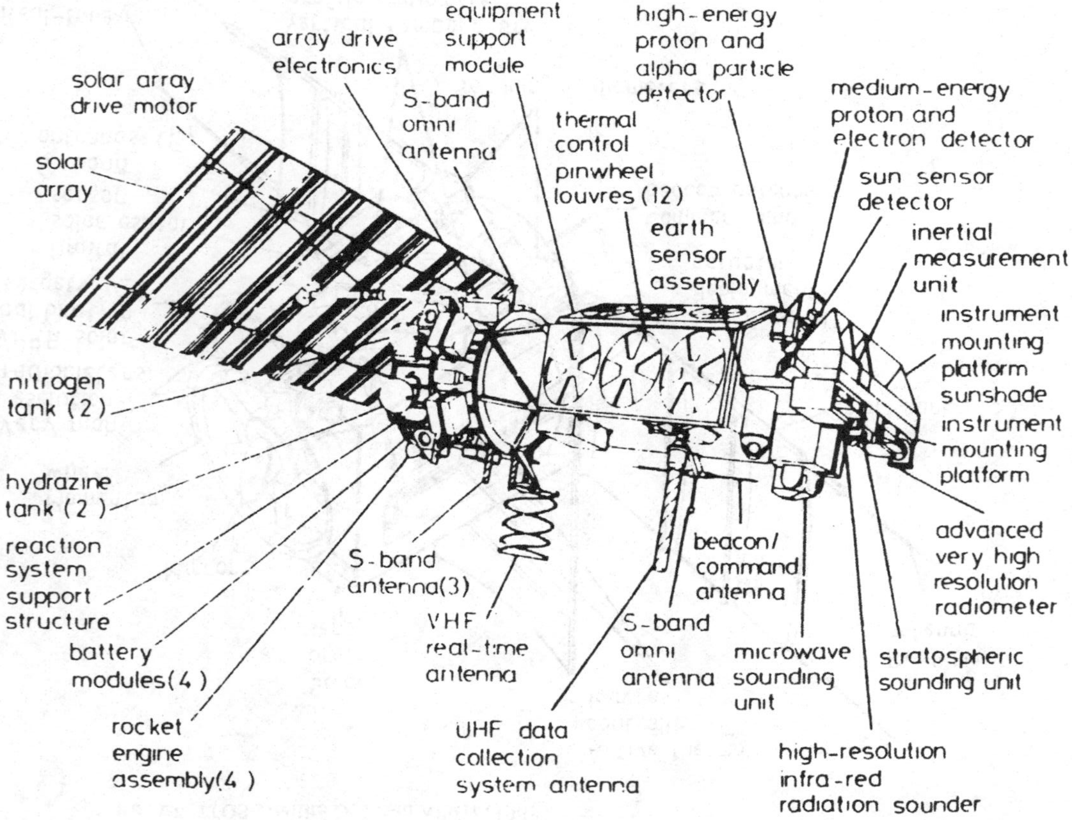

time and a more advanced infrared radiometer for acquiring temperature profiles. These instruments are described further in Ch.6. The orbit of TIROS-N is again sun-synchronous, either a morning orbit crossing the equator at 0730 local time or an afternoon orbit crossing at about 1500 local time. A TIROS-N spacecraft weighs about 740 kg and its power system provides an average of approximately 420 W of power.

Details of the launches of US meteorological satellites are given in Table 2.1.

The Nimbus research satellite series ended with the launch of Nimbus 7 in 1978. The small research satellite Dynamics Explorer (DE) and Solar Mesosphere Explorer (SME) for upper atmosphere research were launched by the US in 1981. Other opportunities for research experiments are being provided by the Space Shuttle. The next major NASA atmospheric mission which is currently

Table 2.1 *American weather satellite launch dates to December 1981*

Satellite	Launch date	Satellite	Launch date
TIROS 1	1 April 1960	Nimbus 3	14 April 1969
TIROS 2	23 November 1960	ITOS 1	23 January 1970
TIROS 3	12 July 1961	Nimbus 4	8 April 1970
TIROS 4	8 February 1962	NOAA 1	11 December 1970
TIROS 5	19 June 1962	NOAA 2	15 October 1972
TIROS 6	19 June 1963	Nimbus 5	11 December 1972
TIROS 7	21 December 1963	NOAA 3	6 November 1973
TIROS 8	21 December 1963	SMS 1	17 May 1974
Nimbus 1	28 August 1964	NOAA 4	15 November 1974
TIROS 9	22 January 1965	SMS 2	6 February 1975
TIROS 10	1 July 1965	Nimbus 6	12 June 1975
ESSA 1	3 February 1966	GOES 1	16 October 1975
ESSA 2	28 February 1966	NOAA 5	29 July 1976
Nimbus 2	15 May 1966	GOES 2	16 June 1977
ESSA 3	2 October 1966	GOES 3	16 June 1978
ATS 1	7 December 1966	TIROS-N	12 October 1978
ESSA 4	26 January 1967	Nimbus 7	24 October 1978
ESSA 5	20 April 1967	NOAA 6	27 June 1979
ATS 3	5 November 1967	NOAA B	29 May 1980
ESSA 6	10 November 1967	GOES 4	9 September 1980
ESSA 7	16 August 1968	GOES 5	22 May 1981
ESSA 8	15 December 1968	NOAA 7	23 June 1981
ESSA 9	26 February 1969		

Table 2.2 *Instrumentation on board operational Meteor 2 satellites*
       (*from Vetlov 1979*)

---

1. Scanning telephotometer for direct transmission of imagery (APT system) in the visible portion of the spectrum (0.5–0.7 μm). Field of view ~2100 km, resolution ~2 km in nadir

2. Scanning infrared radiometer (8–12 μm) with image storage capability. Field of view ~2600 km, resolution ~8 km in nadir

3. TV-type scanning instrument with storage capability for imagery in the visible portion of the spectrum (0.5–0.7 μm). Field of view 2200 km, resolution ~1 km in nadir

4. Scanning eight-channel infrared radiometer (11.10, 13.33, 13.70, 14.24, 14.43, 14.75, 15.02 and 18.70 μm). Field of view ~1000 km, angular resolution ~2°

Fig. 2.6 The Upper Atmosphere Research Satellite (UARS), planned for launch in 1989. For payload see Table 2.4 (from NASA).

planned is the Upper Atmosphere Research Satellite (UARS), particularly
directed at observations of stratospheric structure and composition and due for
launch around 1989 (Fig. 2.6 and Table 2.4).

The Soviet satellite programme is based on the Cosmos series. Those
which are equipped for atmospheric studies with television cameras and infrared
and microwave sensors are given the name Meteor. These spacecraft use inertial-
wheel stabilization to keep the instruments pointing at the surface at all times,
and they occupy polar retrograde orbits like Nimbus.

Operational instrumentation on board Meteor 2 satellites is detailed in
Table 2.2. Research instruments which have been included in Meteor satellite
payloads are shown in Table 2.3. (Vetlov 1979.)

Table 2.3 *Experimental instrumentation on board Meteor satellites*
(*after Vetlov 1979*)

No.	Instruments and characteristics	Application
1.	Spectrometric infrared instruments, 10–17 μm range, angular resolution 6 x 1.5°	Determination of vertical atmospheric temperature profiles
2.	Spectrometer–interferometer developed in the German Democratic Republic, 6.25–25 μm range, angular resolution 2°	Determination of vertical atmospheric temperature profiles and atmospheric water vapour and ozone content
3.	Microwave 0.8 cm polarization radiometer, two orthogonal polarizations, angular resolution 2.5°	Tracking of falling precipitation areas, determination of cloud water content and phase composition
4.	Three-channel microwave 0.8 cm radiometer (scanning in 1000 km band), 0.8, 1.35 and 8.5 wavelengths, resolution (with orbit at a height of 900 km) ~24 x 30 km, 90 x 90 km and 100 x 100 km on the ground, respectively	Determination of the total atmospheric water content, cloud water content, sea-surface temperature; tracking of falling precipitation areas and ice-cover boundaries
5.	Scanning infrared polarimeter, 1.5–1.9 and 2.1–2.5 μm ranges, field of view (with orbit at a height of 900 km) ~2200 km, angular resolution ~3°	Determination of cloud phase composition
6.	Instruments for studying sun–atmosphere relationships, including: four-channel corpuscular spectrometer, 0.3–30 kev range; scanning infrared slant sounding radiometer (0.3–30 μm)	Obtaining information on corpuscular radiation affecting the upper atmosphere, intensity of thermal infrared radiation of the upper atmosphere, which is one of the agents of energy transfer from the upper into the lower atmosphere
7.	Scanning four-channel TV-type instruments (0.5–0.6, 0.6–0.7, 0.7–0.8 and 0.8–1.1 μm), field of view ~1930 km on the ground, resolution ~1000 x 1.700 km in nadir	Obtaining overlapping images of earth's surface and clouds in various ranges of the visible and near-infrared portion of the spectrum
8.	Scanning two-channel TV-type instrumentation (0.5–0.7 and 0.7–1.1 μm), field of view ~1380 km on the ground, resolution ~240 m in nadir	,, ,,

2.2   Geostationary satellites

A satellite placed in an equatorial orbit at an altitude of 36 000 km has a period of 24 hours and so always occupies the same point in the sky relative to an observer on the ground. Simultaneous coverage can thereby be provided over an area of nearly one quarter of the globe. Within this area the motion of clouds and the development of weather systems can be observed continuously. The first meteorological satellite observing from such an orbit was ATS 1 (Advanced Technology Satellite 1) launched in 1966. It consisted of a spin-stabilized cylinder 140 cm long and 150 cm in diameter and carried a spin-scan camera to provide visible images.

After further ATS satellites had been launched, the geostationary satellites became operational in the GOES (Geostationary Operational Environmental Satellite) series (Ensor 1978). In 1977 a European geostationary satellite, Meteosat (Fig. 2.7), carrying visible and infrared imaging devices was launched into position over the equator at the Greenwich meridian. Also in 1977, the Japanese launched GMS (Geostationary Meteorological Satellite) over the western Pacific (see Fig. 3.4).

For the year of the FGGE five geostationary satellites were providing continuous cover of the whole atmosphere up to about 65° latitude (Fig. 2.8).

Each geostationary satellite possesses its own data acquisition and dissemination system. That for Meteosat is shown in Fig. 2.9. Raw data is directed from Meteosat to a central ground station where the data is corrected for geometrical distortions, calibrated and formatted into a variety of formats. It is then retransmitted to Meteosat which relays the processed images to any station with

Table 2.4 *UARS Payload*

WINTERS	Temperature and Wind Measurement in the Mesosphere and Lower Thermosphere
HRDI	High-Resolution Doppler Imager
USSIE	UV Solar Spectral Irradiance Experiment
CLAES	Attitude Distribution of Atmospheric Minor Species and Temperature in the 10 to 60 km range
PEM	Particle Environment Monitor
SUSIM	Solar UV Spectral Irradiance Monitor
HALOE	Halogen Occultation Experiment
ISAMS	Improved Stratospheric and Mesospheric Sounder
MLS	Microwave Limb Sounder
SBUV	Solar Backscatter Ultraviolet Spectral Radiometer
ACRIM	Active Cavity Radiometer Irradiance Monitor

Fig. 2.7 The external appearance of Meteosat. It is about 2m in diameter and weighs approximately 200 kg.

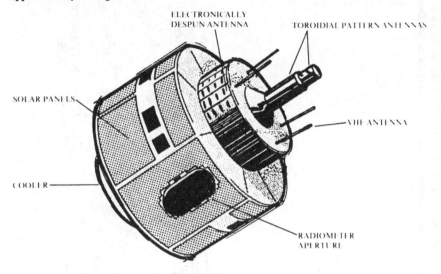

Fig. 2.8 Coverage by the five geostationary satellite operated for the FGGE (WMO 1978). The regions shown are those for which good wind determination can be made (c.f. § 12.2); satisfactory cloud images are obtained over substantially larger areas.

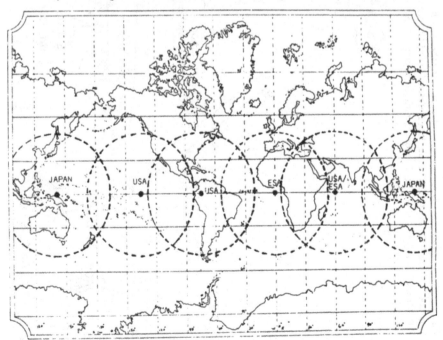

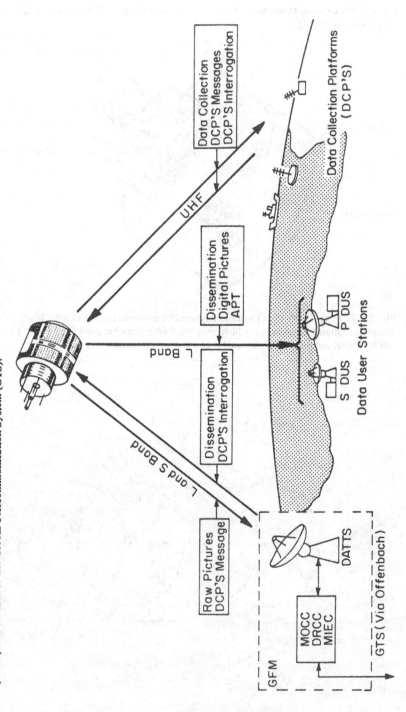

Fig. 2.9 The Meteosat data dissemination system showing the Data Acquisition Telecommand and Tracking Station (DATTS), the Meteosat Operations Control Centre (MOCC), the Data Referencing and Conditioning Centre (DRCC), the Meteorological Information Extraction Centre (MIEC) and the link to the Global Telecommunication System (GTS).

the appropriate reception equipment within its field of view. Data is transmitted in digital form to primary user stations (PDUS) and in analogue form to secondary user stations (SDUS), the latter possessing simpler receivers with analogue facsimile devices at their outputs.

2.3   Missions to the planets

Table 2.5 lists missions to other planets which have flown and one which is currently planned. The most accomplished series of spacecraft are the Mariners of the United States and the Russian Veneras. Fig. 2.10 shows Mariner 9, which became the first man-made satellite of Mars in 1971. Considerable evolution of this design has taken place to produce the latest versions, now called Voyager (Fig. 2.11), the first of which reached Jupiter in 1979 and Saturn in 1980. The most conspicuous difference is the absence of 'solar paddles' from the Voyager spacecraft; the solar radiation intensity in the outer solar system is too low for these devices to be useful. Instead, radioisotope thermal generators are used. These use nuclear energy from the decay of plutonium dioxide to generate 130 W of electricity, at an efficiency of about 6%. The generators are carried at the end of long booms to allow efficient dissipation of excess heat and

Fig. 2.10 Mariner 9 Mars Orbiter spacecraft.

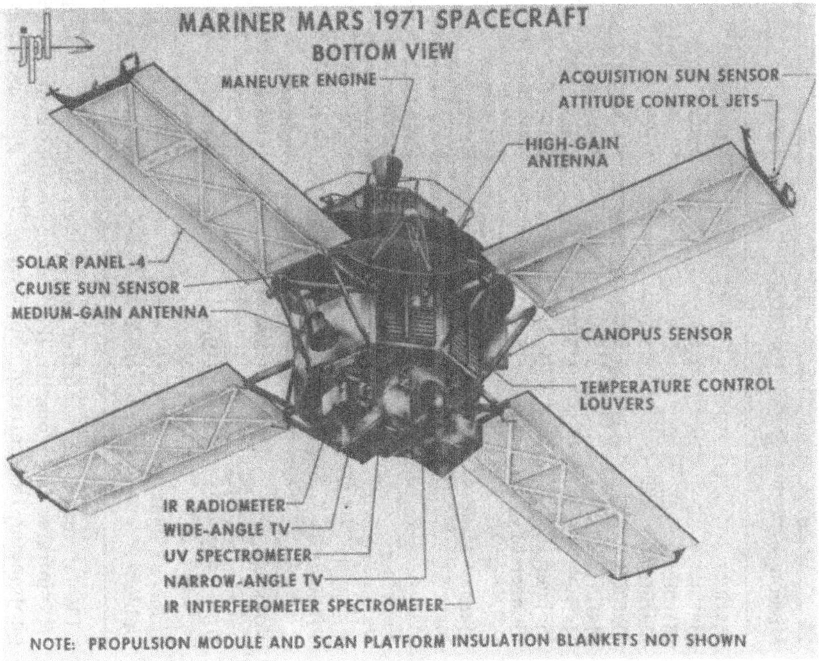

Table 2.5 *Planetary missions using remote sensing*

Mission	Country	Target	Year	Type	Experiments*
Mariner 2	USA	Venus	1962	Fly-by	IRR, MR, RO
Mariner 4	USA	Mars	1965	Fly-by	TV, RO
Mariner 5	USA	Venus	1967	Fly-by	UVS, RO
Mariner 6	USA	Mars	1969	Fly-by	TV, IRR, IRS, UVS, RO
Mariner 7	USA	Mars	1969	Fly-by	TV, IRR, IRS, UVS, RO
Mariner 9	USA	Mars	1971	Orbiter	TV, IRR, IRS, UVS, RO
Mars 3	USSR	Mars	1971	Orbiter	IRR, IRP, VP, MR
Pioneer 10	USA	Jupiter	1973	Fly-by	IP, IRR, UVS, RO
Pioneer 11	USA	Jupiter	1974	Fly-by	IP, IRR, UVS, RO
		Saturn	1978		
Mariner 10	USA	Venus	1974	Fly-by	TV, IRR, UVS, RO
Mars 5	USSR	Mars		Orbiter	IRR, IRS, MR, RO
Mars 6	USSR	Mars		Orbiter	IRR, IRS, MR, RO
Venera 9	USSR	Venus	1975	Orbiter	IRR, IRS, MR, RO, TV, PP
Venera 10	USSR	Venus	1975	Orbiter	IRR, IRS, MR, RO, TV, PP
Viking 1	USA	Mars	1976	Orbiter	IRS, IRR, RO
Viking 2	USA	Mars	1976	Orbiter	IRS, IRR, RO
Voyager 1	USA	Jupiter	1979	Fly-by	IRS, TV, UVS, PP, RO
		Saturn	1981		
Voyager 2	USA	Jupiter	1979	Fly-by	IRS, TV, UVS, PP, RO
		Saturn	1980		
		Uranus	1984		
Pioneer Venus	USA	Venus	1978	Orbiter	IRSR, UVS, IP, RO
Galileo	USA	Jupiter	1987	Orbiter	IRS, UVS, PPR, RO, TV

* TV=television, IRR=infrared radiometer, MR=microwave radiometer, RO=radio occultation, UVS=ultraviolet spectrometer, IRS=infrared spectrometer, IP=imaging photometer, PP= photopolarimeter, IRSR= infrared spectroradiometer, VP=visible photometer, IRP=infrared photometer, PPR= photopolarimeter/radiometer

to reduce the radiation hazard to electronic systems on board the spacecraft. A second obvious difference between the early Venus spacecraft and its latest descendant is the size of the communications antenna; the larger system is needed to communicate over the enormous distance between Earth and Saturn (nearly 1000 million miles).

Mariner spacecraft, like Nimbus, are 'three-axis stabilized', which means that they retain their orientation with respect to a system of celestial co-ordinates using gas jets, star sensors and gyros. Mariner and Nimbus spacecraft

Fig. 2.11 Voyager, an advanced three-axis attitude-stabilized spacecraft for exploration of the outer planets.

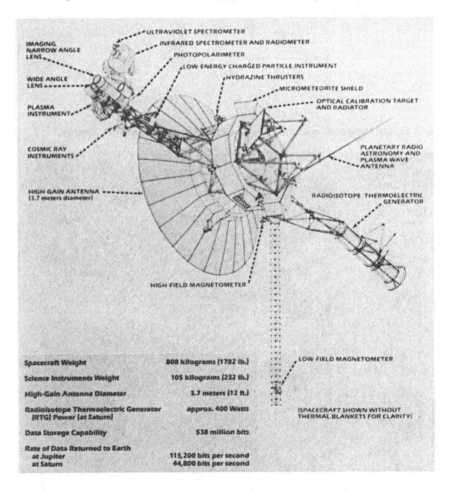

Spacecraft Weight	808 kilograms (1782 lb.)
Science Instruments Weight	105 kilograms (232 lb.)
High-Gain Antenna Diameter	3.7 meters (12 ft.)
Radioisotope Thermoelectric Generator (RTG) Power (at Saturn)	approx. 400 Watts
Data Storage Capability	538 million bits
Rate of Data Returned to Earth at Jupiter at Saturn	115,200 bits per second 44,800 bits per second

are also typically quite similar in size and weight. Mariner 10, for example (Fig. 2.12), weighed 500 kg and consumed about 300 W of power, of which 80 kg and about 60 W were reserved for the scientific instruments.

The other major class of US planetary spacecraft is named Pioneer. The earliest Pioneers were used to explore the interplanetary medium; the first to be used for planetary studies was Pioneer 10 which flew by Jupiter in December 1973 (Fig. 2.13). The latest is the 1978 Venus orbiter (Fig. 2.14).

Pioneer spacecraft are typically smaller and lighter than Mariners. Other than that, the principal difference is the use of spin-stabilization rather than the more complex three axis system. Pioneers rotate at speeds of about 5 rpm, retaining communication with the earth either by placing the spin axis through the axis of symmetry of the antenna towards the earth, as on Pioneers 10 and 11, or by 'despinning' the antennas as on Pioneer Venus.

Fig. 2.12 The Mariner 10 spacecraft, which was the first to visit the planet Mercury.

For the 1976 Viking missions to Mars, an important part of which was the placing of heavy landers on the Martian surface, a new type of orbiter was developed, more sophisticated than either Pioneer or Mariner. This carried two infrared remote sensing instruments and colour television cameras.

Another specially developed spacecraft is Galileo, scheduled to become the first man-made Jupiter orbiter in 1987. This uses the 'dual spin' approach for the first time, whereby half of the module rotates to provide gyroscopic stabilization and to scan the direct-sensing particles-and-fields experiments, while the other half is 'despun', relative to the stars, to provide a stable platform for the remote sensing instruments. Fig. 2.15 shows the design concept for Galileo as it existed some five years before launch.

Until 1975, the planetary exploration programme of the USSR consisted of entry vehicles which made direct measurements of the Venusian atmos-

Fig. 2.13 Pioneers 10 and 11 were identical spin-stablized spacecraft designed for preliminary investigations of the outer solar system. Note the radioisotope power supplies on the large booms.

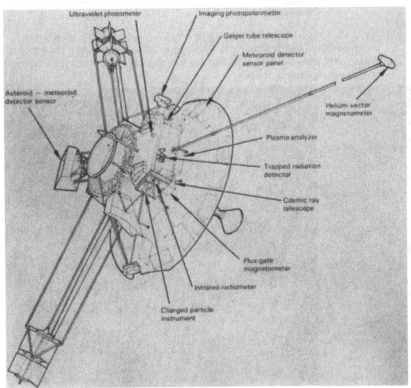

Fig. 2.14 The Pioneer Venus orbiter spacecraft. The dish at the top pointed Continuously at the Earth, while the main drum-like body rotated at 5 rpm to stabilize the spacecraft. The exterior of the drum is covered with solar cells to provide power.

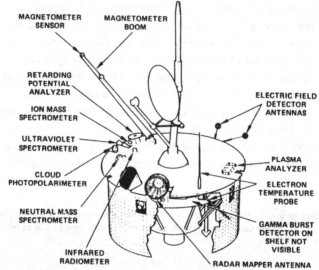

MAGNETOMETER SENSOR
MAGNETOMETER BOOM
RETARDING POTENTIAL ANALYZER
ION MASS SPECTROMETER
ULTRAVIOLET SPECTROMETER
CLOUD PHOTOPOLARIMETER
NEUTRAL MASS SPECTROMETER
INFRARED RADIOMETER
ELECTRIC FIELD DETECTOR ANTENNAS
PLASMA ANALYZER
ELECTRON TEMPERATURE PROBE
GAMMA BURST DETECTOR ON SHELF NOT VISIBLE
RADAR MAPPER ANTENNA

Fig. 2.15 A model of the Galileo Jupiter orbiter spacecraft. The umbrella-like communications dish is deployable. The pointed object on the top is the atmospheric entry probe, with its relay antenna seen on an arm above and the science instrument platform to the left.

Fig. 2.16 The Russian orbiter/lander spacecraft, Venera 10.

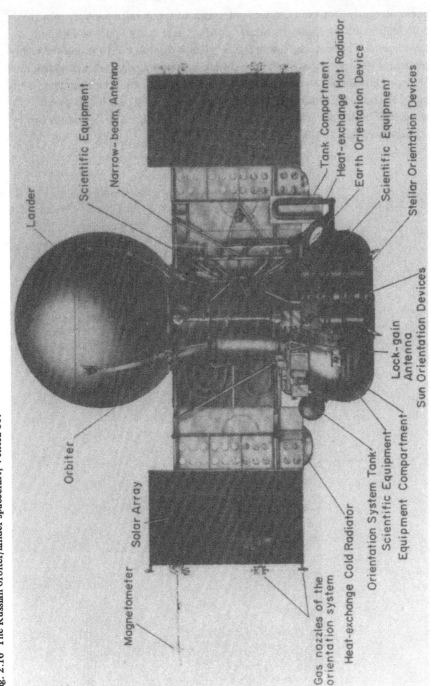

phere. Recently, however, orbiter-lander pairs of spacecraft have been sent to both Mars and Venus, and the orbiter portions have carried remote-sensing experiments. The spacecraft were similar in each case: Fig. 2.16 shows the highly successful Venera 10 spacecraft. The Soviet orbiters are three-axis stabilized, like Mariners, but are not equipped with 'scan platforms' to allow independent pointing of the optical instruments, as the US spacecraft are. The Veneras each weighed 6000 kg in transit, including the lander at about 1500 kg and the orbit insertion motor.

3

VISIBLE, INFRARED AND
MICROWAVE IMAGING

3.1 Visible imagery

3.1.1 Introduction

Terrestrial cloud mapping by photography or television from orbit is a relatively straightforward application of remote sounding. The fractional cloud cover as a function of space and time is an important basic meteorological parameter; also the appearance of the cloud patterns associated with fronts, depressions, storms and other phenomena is characteristic and the correlation between a photograph from a satellite and a weather map is usually substantial (Fig. 3.1). An experienced observer can detect the characteristics of a developing catastrophic system, such as a hurricane, and anticipate its future course; photographs of the same area repeated regularly over a period of time reveal details of the overall circulation system such as cellular patterns and convergence zones. Information from satellite images has been particularly valuable over the Southern Hemisphere where there is a great scarcity of data from conventional sources (Guymer 1978). Purdom (1982) and Liljas (1982) give details of the interpretation of cloud images in terms of mesoscale analyses.

By artificially enhancing the contrast on such pictures, it is possible to pick out areas of very deep cloud with exceptionally high albedo (Suomi 1969) which correlate well with areas of precipitation. In chapter 11 a more detailed discussion of precipitation measurement is given; the use of cloud images for the study of cloud properties and for wind determination is also discussed.

The usefulness of cloud photographs in forecasting applications depends on their being available to those who use them in as near real time as possible. In the early days, because of the difficulty of transmitting photographs quickly over long distances without great loss of clarity and resolution, they were converted first to nephanalyses which represent the important features of the cloud layout in symbolic form.

For later satellites, direct data transmission (known as automatic picture transmission APT) of cloud imagery has been available to relatively simple ground stations. The dissemination arrangements for geostationary satellite pictures have already been mentioned in the previous chapter.

In the following sections brief descriptions of the satellite instrumentation for imaging in different spectral regions will be given together with some examples of the results.

Fig. 3.1 (a) Image including the British Isles and Iceland from the AVHRR on Tiros N at about 1500 Z on 19 May 1979. The occluded front is a region of marked wind shear and exhibits wave structure most likely produced by barotropic instability. Behind the cold front, typical open-celled organisation of the convection is apparent with enhanced narrow bank of cloud prominent near the main frontal cloud mass. (received and processed by the University of Dundee Electronics Laboratory).

3.1.2 *Advanced vidicon camera system (AVCS)*

The first television pictures of the earth were taken from TIROS 1 launched in 1960 with a 500-line vidicon. The system was developed further for the Nimbus 1 and 2 satellites. Called the AVCS it consists of three identical cameras mounted at 35° to each other with the central unit viewing the earth vertically. The field of view of each is arranged to give a slight overlap both between the three units and in successive orbits so that contiguous global coverage results. Each camera has a 1 inch vidicon tube with a 40 ms exposure time; this is scanned by an 800-line raster of 800 points. At the subsatellite point the resolution of objects on the ground is about 1 km. A triplet of photographs is taken every 91 s until the satellite passes over the north pole into the earth's shadow when the system switches off until the satellite sees the sun again approximately 45 min later. Because of the inconstant illumination of the surface, some kind of

Fig. 3.1.(b) Synoptic situation at 1200 Z on 19th May 1979.

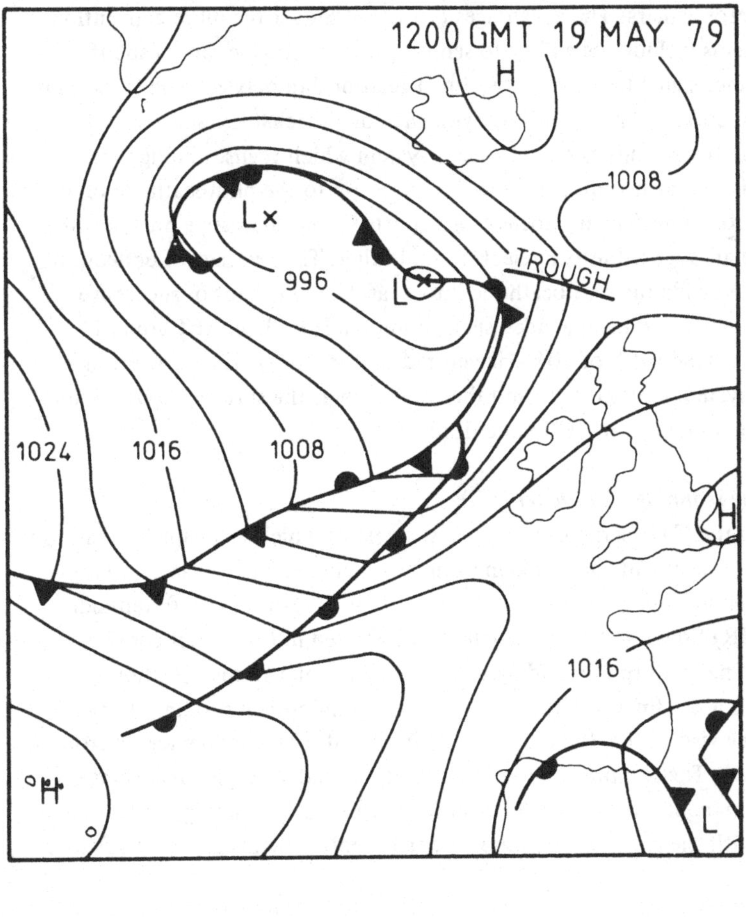

variable aperture stop is needed to prevent over or under exposure at different latitudes. The AVCS uses an iris-type stop which is opened and closed directly by the shaft which carries the solar paddles: these are constrained to keep the solar cells perpendicular to the direction of the sun, so the angular position of the shaft is related to the solar zenith angle at the surface. The phase of the linkage is arranged so that the aperture is wide open near the poles and smallest as the satellite crosses the subsolar point. Calibration is provided by a uniformly illuminated wedge in the field of view of each camera, so that a stripe ranging from black to white through all shades of grey appears down the side of each photograph. The pictures are stored on a tape recorder until a command is received, whereupon up to 200 frames are transmitted to the ground by the 5 W S-band transmitter.

3.1.3   Image dissector camera system (IDCS)

On Nimbus 3 and 4 the three-camera AVCS was replaced by a single image dissector camera. The advantages of the IDCS over the more conventional camera systems include the ability to sense a greater dynamic range (about 100 : 1), higher signal to noise ratios, direct relationship between light flux input and electron current output and the avoidance of a mechanical shutter.

The IDCS consists of a glass lens system which focuses sunlight reflected from the earth's surface, or from clouds, on to the photocathode of an image dissector tube. The photoelectrons emitted from successive portions of the cathode are focused into an electron multiplier. The two sets of deflection coils serve to divide the photocathode into a raster, the output of successive portions of which produces a video signal compatible with an APT ground station (i.e. a bandwidth of 204 kHz centred on 136 MHz). The viewing angle of the lens system is about 108° and the resolution at the surface about 3 km at the nadir for a satellite at 1000 km altitude.

3.1.4   The scanning radiometer (SR)

On the ITOS series of satellites, the first of which was launched early in 1970, a new instrument for visible imaging was included which was a development from an infrared scanning radiometer (the high resolution infrared radiometer HRIR) flown on Nimbus satellites. Illustrated in Fig. 3.2, it consists of a mirror rotating at 48 rpm providing a scan perpendicular to the direction of spacecraft motion. Energy is focused by a Cassegrain optical system into two channels, one visible, 0.5–0.7 μm, and one in the infrared window region from 10.5–12.5 μm. The scanning radiometer is a line-scan device; global coverage is achieved from continuous horizon-to-horizon cross-track scanning by the mirror combined with the forward motion of the spacecraft. The instantaneous field of view at the ground is approximately 4 km for the visible channel and 7.5 km for the infrared channel.

On the later ITOS series beginning with NOAA 2, a further scanning radiometer known as the Very High Resolution Radiometer (VHRR) is included which possesses considerably higher horizontal resolution. The optical design is similar to that shown in Fig. 3.2. To obtain the higher performance for the infrared channel a cadmium–mercury–telluride photoconductive detector is included which is cooled by radiation to space to its operating temperature of ~105 K. The instantaneous field of view at the ground for the VHRR is about 2 km for both channels. An improved APT system called High Resolution Picture Transmission (HRPT) using a broadcast frequency of 1698 MHz is available for data from the VHRR. For the TIROS-N satellite series, four channels have been included in a new VHRR, called the Advanced Very High Resolution Radiometer (AVHRR). One is a visible channel, 0.55–0.68 μm, and one is near infrared, 0.725–1.10 μm. These two channels enable clouds, land-water boundaries, snow and ice extent and ice/snow melt inception to be distinguished. The other two channels are in the thermal infrared and will be mentioned in §3.2. An example of an image from the AVHRR is given in Fig. 3.1.

3.1.5 Geostationary satellite spin–scan camera system

The first pictures of the earth taken from geostationary orbit were from a single channel on ATS 1.

Fig. 3.2 Optical schematic, ITOS scanning radiometer (courtesy of Santa Barbara Research Center) from Schwalb (1972).

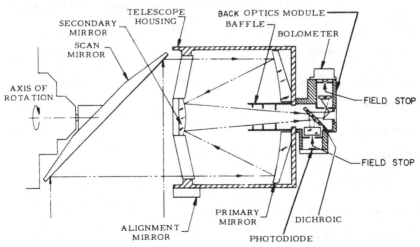

The first routine colour pictures of the earth were taken by ATS 3, launched in November 1967 into a geostationary orbit over the Americas such that simultaneous coverage of some of both the Pacific and Atlantic oceans is obtained. The spacecraft spins at 100 rpm, and this motion is used by the colour camera to scan the earth along lines of constant latitude while a precision step mechanism moves each scan slightly in latitude until, after 24 min and 2400 steps, the whole disk has been covered. The camera itself consists of a high resolution telescope and three photomultipliers, each with a different spectral response corresponding to blue, green and red. The telescope scans while the photomultiplier tubes remain stationary and are fed by fibre optics from separate apertures in the focal plane. The instantaneous field of view of the telescope is equivalent to about 4 km of resolution at the surface. The colour signals are transmitted to the ground instantaneously, since storage is not required on a geostationary satellite, and the picture is assembled at the receiving station.

As described in Ch.1, geostationary satellites relay formatted images to user stations within less than an hour of their being taken.

The Meteosat radiometer which includes two visible and two infrared channels (Table 3.1) is illustrated in Fig. 3.3. Scanning is carried out in a similar manner to the ATS spin-scan camera.

Table 3.1 *Meteosat radiometer channels*

	Visible	IR (water vapour)	IR (normal)
Spectral bands	0.4–1.1 μ	5.7–7.1 μ	10.5–12.5 μ
Number of channels	2 (simultaneous)	1 (in time sharing with 1 vis channel)	1 (+1 redundant)
Number of lines/pict.	5000 (2500)*	2500	2500
Number of samples/line	5000	2500	2500
Resolution (sub sat. pt.)	2.5 km	5 km	5 km
Line duration	30 ms		
Line recurrence	600 ms		
Image taking duration	25 min		
Image recurrence	30 min		
Transmission → DATTS	DIGITAL 166 Kb/s (normal) 2.7 Mb/s (back up)		

*In brackets: case where water vapour channel is also transmitted

Fig. 3.3 Meteosat radiometer layout (from European Space Agency).

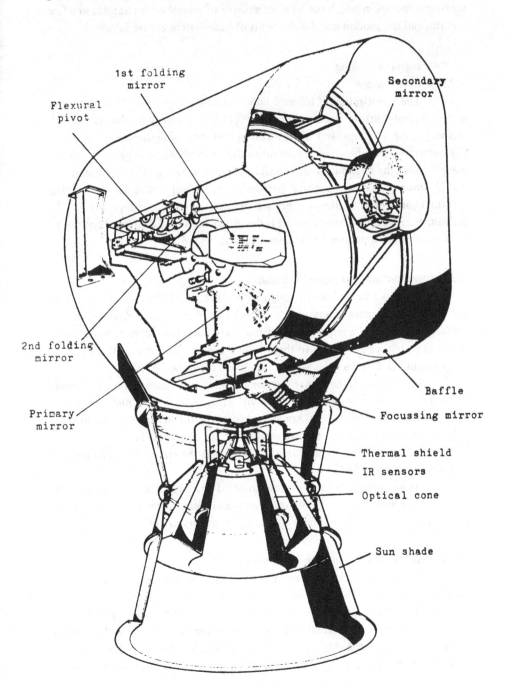

Images from geostationary satellites can be strung together to produce time-lapse movies in which the weather activity of several weeks unfolds in a few minutes and the motion and development of each system can be followed.

3.2   Infrared imaging
3.2.1   Introduction
The investigation of infrared radiation emitted by the earth-atmosphere system is important for three main reasons: (1) loss of energy by the system in this way is the main heat sink for the atmospheric thermodynamic engine; (2) investigation of the spectral content of the emitted radiation can give information about the distribution of atmospheric temperature and composition; and (3) investigation of the spatial distribution of the emitted radiation in certain spectral regions enables clouds or surface characteristics to be mapped. In 3.2.2 instrumentation for infrared imaging is described and in 3.2.3 some of the problems of surface-temperature measurement are indicated.

3.2.2   Instruments for infrared imaging
Nimbus 1 carried the first high (spatial) resolution instrument for infrared imaging from space. Known as the High Resolution Infrared Radiometer (HRIR), it contained a single channel from 3.4–4.1 μm in an optical layout and with a scanning system similar to that of the scanning radiometer (SR) already described in 3.1.4. In fact the SR was developed from the HRIR.

On Nimbus 4, the longer wavelength window channel at 10–12 μm was utilized so that observations could be taken during both day and night, with a resolution of 7 km from a 1000 km altitude orbit. A further spectral channel 6.5–7.0 μm was included to observe emission in the centre of the water vapour band (§10.2). The resulting instrument is known as the temperature and humidity infrared radiometer (THIR).

As has already been mentioned infrared channels have been included on geostationary satellites (§2.2) and on operational meteorological satellites. In Fig. 3.4 is illustrated an infrared image from a geostationary satellite.

3.2.3   Surface-temperature measurement
The temperature of the surface is clearly an important meteorological quantity; it is one of the parameters which determines the amount of heat flow between the surface and the atmosphere. In the case of the land the surface temperature varies very considerably with the type of surface and also with the time of day (see, for instance, Chen *et al.* 1979). At the same time the surface temperature can be very considerably different from the temperature of the atmos-

phere near the surface or the 'screen' temperature, which is that normally
employed by meteorologists.

The same problems, however, are not present to the same degree for
the sea surface, so that more precise measurements are both possible and more
meaningful. Since the amount of heat and moisture taken up by the atmosphere
during its passage over the oceans is very dependent on the surface temperature,
it is perhaps not surprising that correlations appear to exist between the climatic
type and its change from year to year and anomalies which may only be a few
degrees Celsius in the temperature of parts of the ocean surface. Accurate
measurement of sea-surface temperature is, therefore, of considerable
importance–an accuracy of 0.25 °C as a suitable aim has been quoted as one of
the requirements for the GARP.

Fig. 3.4 Infrared image from the Japanese Global Meteorological Satellite (GMS-1)
taken at 0300 GMT on 12 October 1979.

Two main spectral regions ('windows') are available in the infrared where the atmosphere is substantially transparent, 3.4–4.1 μm and 10.5–12.5 μm. On all satellites prior to Nimbus 3, infrared imaging systems used the first of these intervals. Its chief disadvantage is that some reflected solar radiation is present, in an amount which depends on the nature of the underlying surface.

The radiance L_ν observed in one of these spectral intervals under cloud-free conditions is approximately

$$L_\nu = \epsilon_\nu \tau_\nu B_\nu(T_s) + (1 - \tau_\nu) B_\nu(T_a) \tag{3.1}$$

where $B_\nu(T)$ is the Planck function appropriate to the frequency interval and the temperature T, and τ_ν is the transmission of the atmospheric path between the surface and the satellite instrument, averaged over the spectral interval of observation. The second term in equation (3.1) describes emission from the atmosphere, T_a being an average atmospheric temperature found for any particular case by integration of the radiative transfer equation (§5.3). For conditions where the surface emissivity ϵ_ν is appreciably less than unity, a further term must be added to equation (3.1) describing atmospheric emission reflected from the surface into the path of observation. In order to determine T_s, therefore, it is necessary to know the emissivity ϵ of the surface, details of the absorption by the atmosphere, the atmospheric temperature structure, especially near the surface, and whether there are any clouds present or not. We shall deal with each of these questions in turn.

For parts of the earth well covered with vegetation the emissivity is approximately unity in both spectral regions. Other types of surface, notably rock and sand, can have emissivities as small as 0.5. The emissivity of the ocean surface is very close to unity throughout the region.

Atmospheric absorption in both intervals is due to carbon dioxide and water vapour. Collision-induced absorption by nitrogen (Farmer and Houghton 1966) also occurs in the 3.5–4.1 μm region. Correction for CO_2 absorption can be made as its abundance and distribution are known. In general, however, water-vapour content is not known accurately. Because water-vapour content and temperature are closely related, Smith *et al.* (1970) have found it adequate for the 3.5–4.0 μm region to employ a simple expression for the correction ΔT due to atmospheric absorption which is dependent only on observed brightness temperature T_B (in kelvin) and viewing angle θ (in degrees) from the nadir

$$\Delta T = \left\{ a_0 + a_1 \left| \frac{\theta}{60} \right|^{a_2} \right\} \ln \left(\frac{100}{310 - T_B} \right) \tag{3.2}$$

For $210\,K \leqslant T_B \leqslant 300\,K$ and $\theta \leqslant 60°$, with $a_0 = 1.13$, $a_1 = 0.82$, $a_2 = 2.48$. For $\theta = 0$, ΔT is less than 3 K for very warm surface temperatures.

The correction for absorption in the 10.5–12.5 μm region can be considerably larger than this. Our knowledge of the atmospheric absorption in this region is still inadequate for the accurate measurements required. Because of a contribution from water dimer molecules, that is $(H_2O)_2$, an absorption coefficient k of the form (Bignell 1970)

$$k = k_0 + k_1 e \tag{3.3}$$

must be employed where e is the water-vapour pressure in atmospheres. The value of k_0 appears to be small and a suitable value for k_1 (Burch 1970, Houghton and Lee 1972) is 11.5 g^{-1} cm^2 atm^{-1} at 288 K. k_1 is rather strongly temperature dependent and varies approximately as $\exp(1745/T)$ with T in kelvins. The value of the correction ΔT may amount to 5 K or more for a warm humid atmosphere; so that rather accurate knowledge of the water-vapour content and atmospheric temperature distribution is required if the correction is to be made properly.

In order to make such corrections, radiance measurements from several different window regions may be compared. In addition to the two window regions mentioned above, measurements may be made near 2.3 μm wavelength where the atmosphere's transparency is high and where the Planck black-body function varies very rapidly with temperature (Fig. 3.5). This 2.3 μm channel was included in the Selective Chopper Radiometer (SCR) on the Nimbus 5 satellite (§6.7). Fig. 3.6 illustrates measurements from all three window channels over the tropical Pacific. Estimates of corrections which need to be made to the equivalent temperature, arising from the atmospheric contribution to the emission, for the clear region near 11°N shown in Fig. 3.6 are 1.0 K, 2.6 K and 7.0 K for the 2.3 μm, 3.5 μm and 11 μm channels respectively (Houghton 1979).

Information regarding atmospheric transparency may also be obtained by comparing radiances measured at different frequencies in the 10–12 μm window region, within which water-vapour absorption varies considerably (Prabhakara *et al.* 1974, 1979). For this purpose, later satellites in the TIROS-N series will carry versions of the AVHRR (§3.1.4) with three thermal infrared channels, respectively 3.55–3.93 μm, 10.3–11.3 μm and 11.5–12.5 μm. Further information regarding operational procedures for sea-surface temperature determination is given in Hawson *et al.* (1979).

Another method of tackling the problem which should lead to considerably increased accuracy of sea surface temperature determination, especially in tropical regions, is to use measurements of radiance of the same part of the sea surface but made at different angles to the nadir and therefore including different atmospheric paths – a technique which is employed by the Along Track Scanning Radiometer (ATSR) being built for the European Space Agency's ERS-1 satellite (Fig. 3.7). Comparison of the observations at the different angles enable correction for water vapour absorption and emission to be made. An accuracy of better than ±0.5 K in sea surface temperature determination is claimed for this instrument. (Harries *et al.* 1983.)

There remains the problem of clouds interfering with the path of radiation from the surface to the satellite. A method to deal with this has been developed by Smith *et al.* (1970) who have analysed large amounts of data from the scanning radiometer on Nimbus 2. Because the spatial variability of sea-surface temperature is small it is possible to select from a histogram of brightness temperatures taken over a period of time, values that are high enough and also that recur often enough to be characteristic of the surface.

Fig. 3.6 illustrates another potential method of identifying regions of broken cloud by combining measurements from several infrared window channels where the variation of Planck function with temperature is very different. Further information about cloud cover can be provided during the day from images at visible wavelengths.

Fig. 3.5 Plots of effective temperature against radiance for the three window channels of the Nimbus 5 SCR (see para 6.7), namely D3 (2.3 μm), D4 (3.5 μm) and C4 (11μm). The abscissa gives the scale for C4. The radiances for the channels D3 and D4 have been scaled so as to be equal of C4 at 293K.

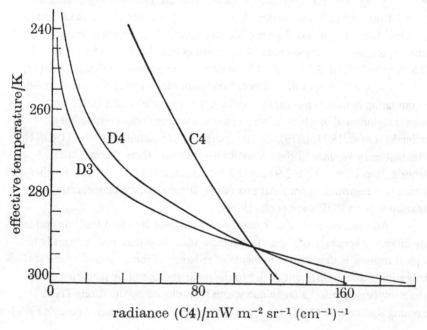

3.3   Microwave imagery

The Nimbus 5 and Nimbus 6 satellites carried an electrically scanned microwave radiometer (ESMR). This consists of a stationary slotted waveguide array about 1 m square with a sensitive mixed receiver operating at a frequency of 19.35 GHz (1.55 cm wavelength) on Nimbus 5 and 37 GHz (0.81 cm) on Nimbus 6. Electronic techniques are employed to sweep the beam from side to side across the satellite's track over the earth's surface so that no moving parts are needed. In each such sweep, the beam is halted (by an onboard computer) at 78 positions for 47 ms each, while a brightness temperature is observed with an accuracy of 1.5 K. Four seconds are required for each sweep.

Fig. 3.6 Observations in three window channels at wavelengths of 11 μm, 3.5 μm and 2.3 μm respectively, of the Nimbus 5 SCR on December 17 1972. Radiance is plotted as equivalent temperature (from Houghton 1979).

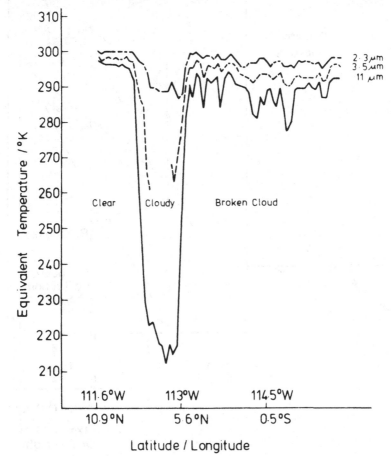

At this wavelength the atmosphere is substantially transparent in the absence of precipitating clouds. The brightness temperature is a measure both of surface temperature and surface albedo. The most striking feature of ESMR images (Webster *et al.* 1975) is the strong contrast between land and water mainly due to the very low emissivity of water at microwave wavelengths. Brighter regions over water in the intra-tropical convergence zone near the

Fig. 3.7 The Along Track Scanning Radiometer for the European Space Agency's ERS-1 satellite (from Harries *et al.* 1983).

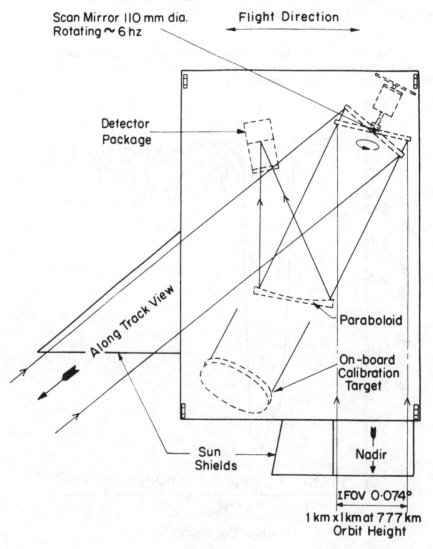

Scan Mirror 110 mm dia. Rotating ∼ 6 hz

Flight Direction

Detector Package

Along Track View

Paraboloid

On-board Calibration Target

Sun Shields

Nadir

IFOV 0·074°

1 km x 1 km at 777 km Orbit Height

equator indicate the presence of very high humidity and precipitation (§11.3).
Allison *et al.* (1979) show how ESMR data has been employed in a study of
floods in Australia in 1974.

Because ice possesses a much higher emissivity than water it is easy to
distinguish the presence of ice in the polar regions. Fig. 3.8 illustrates the appli-

Fig. 3.8. Sea ice cover during different years in the Antarctic region from Nimbus 5 ESMR
data. The area of open water within the ice pack is shown, also the area of ice and the area
of ocean with more than 15% ice cover. Note the progressive decrease in ice cover during
the 4 years (from NASA).

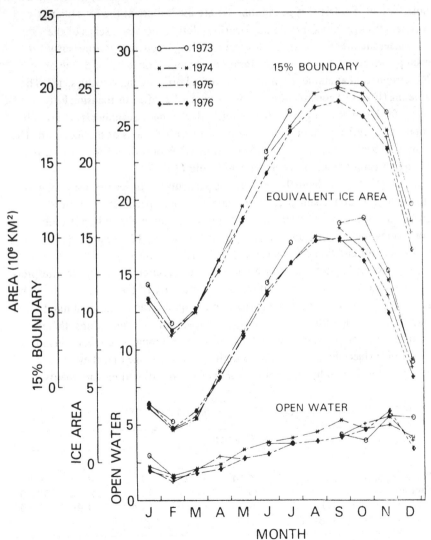

cation of ESMR data to ice cover studies. Also new ice may be distinguished from old ice through its higher emissivity (Webster *et al.* 1975, Gloersen *et al.* 1974). Because, also, the emissivity of the land surface is a strong function of its moisture content the possibility exists (Schmugge 1976, Burke *et al.* 1979) of measuring soil moisture content – an important quantity for both meteorological and hydrological studies.

On the Nimbus 7 satellite and on SEASAT, both launched in 1978, the Scanning Multichannel Microwave Radiometer (SMMR) was mounted. This is a ten-channel instrument measuring at five wavelengths (Table 3.2) in orthogonal polarisations. Six Dicke-type radiometers are used, those at the four longest wavelengths measure alternate polarizations during successive scans of the antennae, the other two, at the shortest wavelength, operate continuously for each polarization. A two-point calibration system is provided consisting of an RF termination at ambient temperature and a horn antenna viewing space. The antenna (Fig. 3.9) consists of a 42° offset parabolic reflector focusing into a single feedhorn covering all of the wavelength channels. Scanning is achieved by rotating the reflector about an axis coincident with the axis of the feedhorn. The Nimbus 7 SMMR weighs 52 kg and consumes 60 W of power. Further instrumental details can be found in Gloerson and Hardis (1978).

Of the five channels, channel 1 is particularly for sea-surface temperature determination, channel 2 for sea-state and wind-speed determination, channels 3, 4 and 5 for water vapour, liquid water and precipitation. Results concerning these latter quantities will be presented in §12.2 (wind), §10.2 (water vapour and liquid water) and §11.3 (precipitation).

Regarding the sea-surface temperature measurement, accurate interpretation is complicated by the variation of the emissivity of the sea surface with sea state (Fig. 3.10). Further, some correction is required for attenuation due to water vapour. Algorithms for sea-surface temperature determinations, therefore, require data from at least three SMMR channels. Although microwave measurements of surface characteristics possess the great advantage that they are relatively unaffected by cloud cover compared with infrared measurements,

Table 3.2 *SMMR channels (after Gloerson and Hardis 1978)*

	Channel				
	1	2	3	4	5
Wavelength (cm)	4.54	2.80	1.66	1.36	0.81
Frequency (GHz)	6.60	10.69	18.00	21.00	37.00
RF bandwidth (MHz)	4.20	2.60	1.60	1.40	0.80
Antenna beam width (deg)	25°	25°	25°	25°	25°

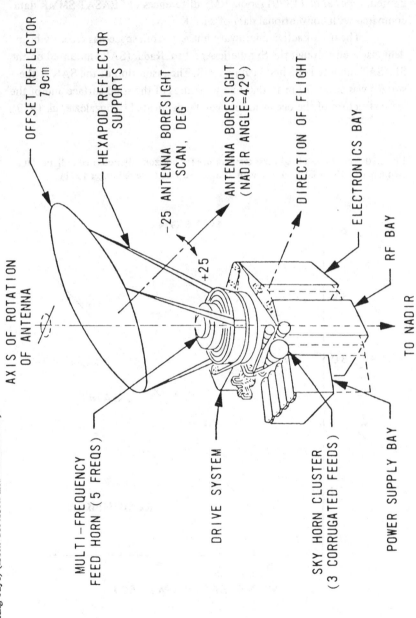

Fig. 3.9 SMMR instrument configuration showing antenna, feed horn drive assembly, and electronic boxes (configured to fit the Nimbus 7 sensor ring bays) (from Gloerson and Hardis 1978).

because of the corrections required, particularly for sea state, the accuracy of sea-surface temperature determination by microwave radiometry is severely limited. Lipes *et al.* (1979) report RMS differences of SEASAT SMMR data compared with conventional data of ±1.5 K.

The use of active microwave imagery from space has recently been demonstrated through the Synthetic Aperture Radar (SAR) mounted on the SEASAT satellite launched in June 1978. The resolution of the SAR at the surface was about 25 m so that many features of the solid surface and of the wave structure of the ocean surface can be delineated (Gonzalez *et al.* 1979).

Fig. 3.10 Variation of brightness temperature for horizontally polarised radiation at a frequency of 19.34 GHz with roughness of sea surface (after Hillinger 1971).

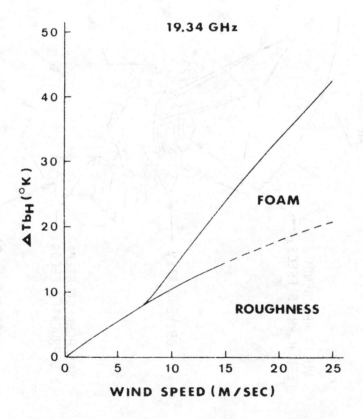

4

RADIATION BUDGET MEASUREMENTS

4.1　Components of the earth's energy budget

The only important source of energy input to the earth-atmosphere system is the sun, this input being balanced on the average by emission of infrared radiation to space. The emitted energy, however, does not have an identical distribution either in space or time with that absorbed from the sun – redistribution and some storage of the energy occur within the atmosphere and the oceans. A detailed discussion of atmospheric energetics is presented in §4.6. A very important parameter in that discussion is the distribution at the top of the atmosphere of the net radiation, i.e. the difference between the incoming solar radiation and the sum of the solar radiation reflected or scattered back to space and the emitted long-wave radiation. Before giving further details of radiation budget studies, some of the instruments which have been used for radiation budget measurements will be described and some results from them introduced.

4.2　The Wisconsin radiometers

Comparison of the emission from a black body for solar and terrestrial temperatures (Fig. 4.1) shows that reflected solar radiation and emitted terrestrial radiation occupy distinct wavelength regions with virtually all radiation with $\lambda > 4.5\,\mu$m being terrestrial and all that with $\lambda < 4.5\,\mu$m being solar. The spectral selectivity required to separate the two is, therefore, minimal. For their pioneering radiometer, first launched on Explorer VII in 1959, Suomi and his co-workers achieved this kind of spectral selectivity simply by using different kinds of paint. The instrument (Fig. 4.2) consisted of two hemispherical infrared detectors, mounted on mirrors to increase their light gathering power, one of which was painted white to reflect all wavelengths shorter than 4.5 μm while the other was matt black and absorbed all wavelengths. Viewing from horizon to horizon at satellite altitude the sensors are self-integrating with greater weight being given to energy arising from the nadir. Aside from their simplicity and ruggedness the most worthy feature of the Wisconsin sensors is their ability to self-calibrate against direct solar radiation twice each orbit as the satellite passes from day to night. This allows a check of prelaunch calibration and continuous compensation for any drift or degradation of the sensors during their experimental lifetime.

A modified instrument known as the flat plate radiometer has been mounted on the ITOS series of satellites. It includes two pairs of sensors, one pair being very similar in performance to those on the TIROS satellites, the other pair being of a thermal feedback design in which the energy required to maintain a constant temperature is measured. This second pair, therefore,

Fig. 4.1 Black-body curves for solar and terrestrial temperatures. Solar radiation absorbed by the earth atmosphere system is, on average, balanced by emission of infrared radiation to space.

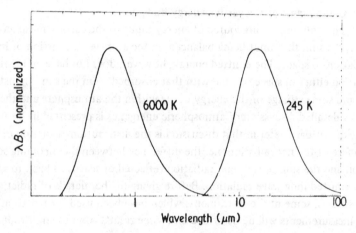

Fig. 4.2 Schematic diagram of Wisconsin wide-field radiometer (from Astheimer *et al.* 1961).

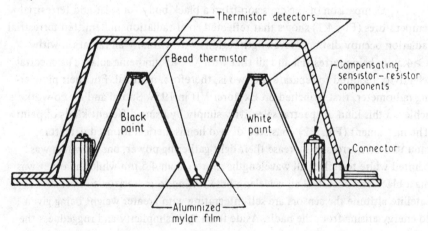

measures the input radiant energy directly apart from a correction due to losses to the mounts. This correction can be determined before launch and also from inflight calibration which is carried out by arranging for the instrument to view its housing from time to time.

4.3   The Medium Resolution Infrared Radiometer (MRIR)

The TIROS medium resolution infrared radiometer has five channels, identical except for the filters. One is shown diagrammatically in Fig. 4.3.

The field of view of the radiometer is approximately a 5° cone, so that the horizontal resolution when viewing the nadir is about 80 km. The earth was scanned by the radiometer which was mounted at 45° from the satellite's spin axis (Fig. 4.4). The channels are located as shown in Fig. 4.5. Measurements from channel 1 have been interpreted in terms of water-vapour content of the upper troposphere (§10.2). The atmospheric window channel, number 5, measures the same spectral interval as that on the THIR (§3.2.2) and measures either cloudtop or surface temperature when each is in the field of view. Since cloudtops are invariably much cooler than the surface below, these measurements were used to produce, for the first time, rather crude maps of the clouds on the night side of the earth. Channel 2 was designed to accept the same wavelengths as the television system, for correlation studies. Channels 3 and 4 measure reflected solar and emitted thermal radiation respectively; it is these which enable estimates of net radiation to be obtained.

At any one time the radiometer measures the intensity of radiation along a specific direction as modified by the spectral characteristics of the

Fig. 4.3 Schematic diagram of medium resolution infrared radiometer. One half of the chopper disc is reflecting and the other half black.

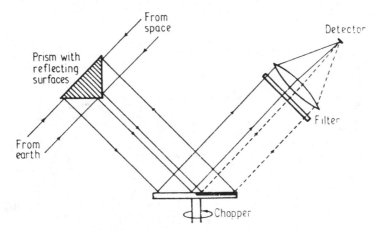

Fig. 4.4. Location and orientation of radiometers in TIROS satellite (from Astheimer *et al.* 1961).

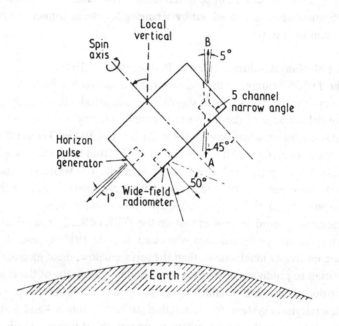

Fig. 4.5 Overall spectral transmission of optical elements in the five channels of the MRIR (from Astheimer *et al.* 1961).

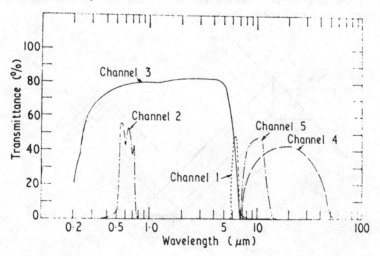

filters. To determine the net radiation, knowledge of the two radiation *fluxes* is required, that is, the radiation leaving unit area of the atmosphere integrated over all directions, whereas what is measured is the radiation in one direction only. Wark *et al.* (1962) and Lienesch and Wark (1967), using in the one case model atmospheres and in the other MRIR measurements themselves, have calculated the angular dependence of thermal radiation leaving the atmosphere under different conditions and have also shown how correction can be made for the significant radiation outside the pass band of the filter so that the outgoing flux of thermal radiation can be obtained from measurements in channel 4.

Similar problems are encountered with interpretation of the reflected solar radiation in channel 3. Sunlight reflected from clouds is not isotropic; in general clouds reflect much more strongly for low zenith angles of the sun. Since, over a period of time, measurements are made with the TIROS radiometer over a wide range of solar zenith angles and directions of view of the radiometer, it is possible to build up information about the angular distribution of reflected solar radiation from the measurements themselves (Raschke *et al.* 1968, Ruff *et al.* 1968). Corrections may then be applied to individual measurements to estimate the local albedo.

4.4   Monitor of Ultraviolet Solar Energy (MUSE) experiment

There has always been a great deal of interest in the possibility of changes with time in the amount of incoming solar energy and the resulting effect these might have on the climate. Measurements of possible variations in the solar flux have concentrated, in the first place, on the ultraviolet region. Although only about 3% of the solar energy lies at wavelengths shorter than 0.3 μm, because of strong oxygen and ozone absorption the energy budget of the mesosphere and stratosphere is very sensitive to this ultraviolet solar energy. Also, it is more likely that variations in solar activity will occur at short wavelengths where the radiation originates in the outer layers of the solar atmosphere.

The MUSE experiment on Nimbus 3 was devised to monitor solar flux in three spectral bands between 0.115 μm and 0.3 μm; on Nimbus 4 five spectral bands were monitored by a similar instrument. The sensors are vacuum photodiodes coupled with appropriate interference filters and are calibrated before launch (Heath and Westcott 1970).

Variations of solar flux with the 27-day period of solar rotation have been observed at the shorter wavelengths and are typically ~5% near 0.18 μm and decrease rapidly with increasing wavelength (Heath 1973). Evidence for variation with the 11-year solar cycle has also been found.

4.5   The Earth Radiation Budget (ERB) experiment

A more complete and accurate instrument for radiation budget measurements, the ERB experiment, has been mounted on the Nimbus 6 and 7 satellites. Specific objectives of the experiment (Smith *et al.* 1977) are to (1) measure incoming solar radiation in total (the solar constant) and in spectral subdivisions of the ultraviolet and visible spectrum; (2) measure earth radiation flux at satellite altitude in the short-wave region (0.2-3.8 μm) and in total ($\lambda > 0.2 \mu$m) from which the planetary albedo, total long-wave flux and net radiation flux can be accurately specified for the time of the Nimbus orbit; (3) measure solar flux and earth reflected short-wave flux for $\lambda \leqslant 0.7 \mu$m and $\lambda \geqslant 0.7 \mu$m to assess the contribution of aerosols to any detectable variations of the earth's planetary albedo; (4) measure earth-reflected solar radiation (0.2-4.7 μm) and earth-emitting long-wave radiation (5-50 μm) with narrow-angle scanning channels in order to model the angular characteristics of the out-going radiance as a function of surface and atmospheric condition (especially cloudiness) and to provide direct measurements of the radiation budget on regional scales (500 km) on a monthly mean basis for climate monitoring and prediction.

Ten channels are included in the Nimbus 6 ERB instrument for measurements directly on the sun; their spectral characteristics are shown in Fig. 4.6. Thermopile detectors are used in these channels which include no imaging optics, only filters, windows and apertures. Channels 1 and 2 are identical except that channel 1 is normally shuttered; from time to time the shutter is removed to provide an inflight check of possible degradation to channel 2. With these channels it has been established that no variations in the solar constant greater than ~0.1% occurred during the first nine months of operation and that variation in any of the solar channels is less than 1% over the same period (Smith *et al.* 1977, Jacobowitz *et al.* 1979). However, the Nimbus 7 ERB instrument includes a cavity pyrheliometer for monitoring the solar constant. Preliminary analysis of the first 1000 days of this data indicate a secular decrease of about 0.3 W m^{-2} per year, and an r.m.s. variability of about 1 W m^{-2} which appears to be related to sunspot activity (Hickey *et al.* 1980). More recent measurements of variations in the solar constant are shown in Fig. 4.7. Earth-emitted infrared radiation and earth-reflected solar radiation are measured with wide-angle field of view channels which view the entire earth disk from satellite altitude. Two channels (again one shuttered for inflight checks) have no filters or windows but are sensitive to all radiation over the wavelength range 0.2-50 μm. Other channels measure only in the regions 0.2-3.8 μm and 0.7-2.8 μm.

The ERB experiment also includes narrow-angle scanning channels to measure the angular distribution of the emerging radiation. Fig. 4.8 illustrates

Fig. 4.6 Spectral intervals monitored by the ERB solar channels (with 1971 NASA standard extraterrestrial solar curve).

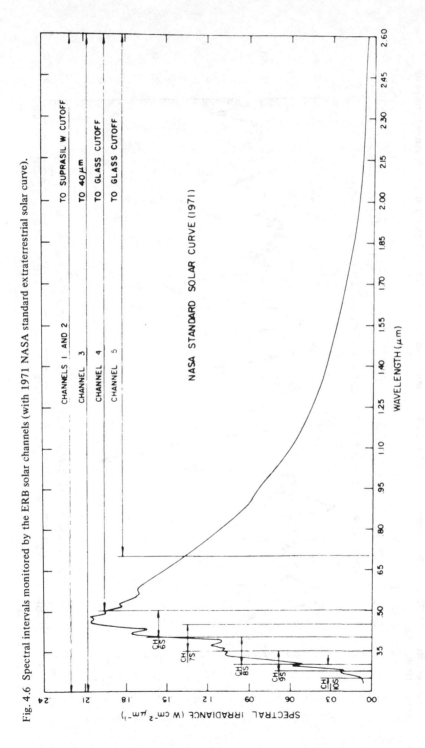

Fig. 4.7 Total solar irradiance as measured on the Solar Maximum Mission shown as a percentage variation about the weighted mean for the first 153 days of the mission. The instrument employed was the Active Cavity Radiometer Irradiance Monitor (ACRIM) (Willson (1980) (Willson (1980) and Willson and Hudson (1981). The large decreases of 0.1% to 0.2% are clearly correlated with the development of specific sunspot groups on the sun (from NASA).

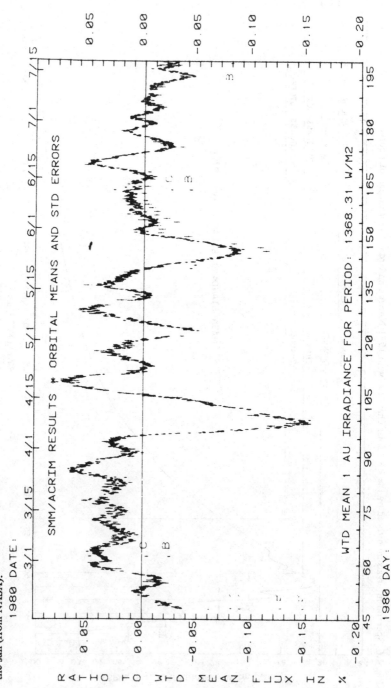

that to use measurements at satellite altitude, typically 1000 km, to determine the outgoing flux from local areas of the earth, the angular distribution of out-going radiation must be known. Typical measurements of reflection and emission as functions of angle are shown in Fig. 4.9.

4.6   Results and studies of atmospheric energetics

Vonder Haar and Suomi (1971) have summarized the results of radiation balance studies over the period 1962–6 obtained from both TIROS and Nimbus satellites and from both types of sensor described in §4.2 and §4.3. The results they obtained from year to year are very consistent although they point out that, because the satellite measurements are not completely representative over different times of day, there may be a bias in the results arising from a possible diurnal variation of the radiation budget. Fig. 4.10 shows mean meridi-onal profiles of components of the earth's radiation budget for 1962–6. The average albedo for the period is about 0.3, rather less than had previously been estimated. They also found that each hemisphere has nearly the same planetary albedo and infrared loss to space on the mean annual time scale. This points out the dominant influence of clouds on the energy exchange with space since the surface features of the two hemispheres are quite different.

Fig. 4.8. Illustrating the problem of relating the outgoing radiative flux at the top of the atmosphere to measurements at a typical satellite altitude.

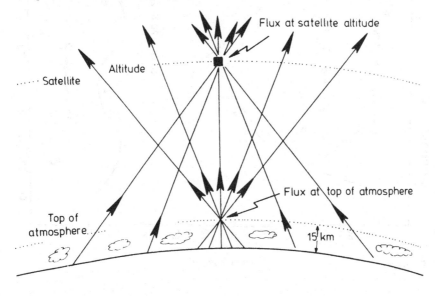

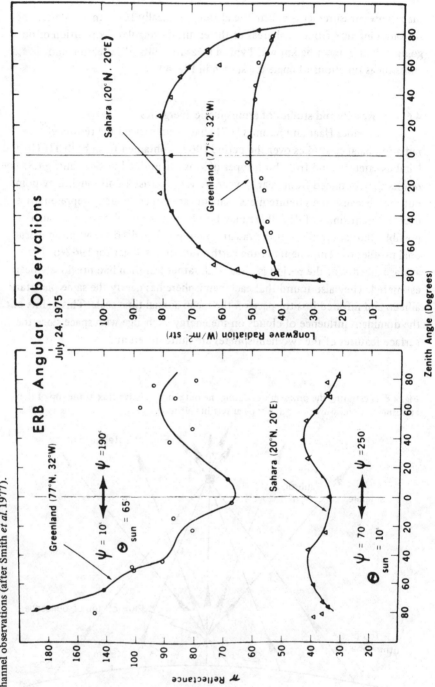

Fig. 4.9 Angular reflectance observations over Greenland and the Sahara Desert, 24 July 1975 from Nimbus 6 ERB narrow angle scanning channel observations (after Smith *et al.* 1977).

Regional studies of the net energy exchange at the top of the atmosphere are also important. Fig. 4.11, for instance, illustrates the large radiative 'sink' over the Sahara desert caused by the sand surface being both hot (leading to high long-wave radiation) and highly reflective (leading to high reflected solar radiation). A recent summary of radiation budget information has been given by Stephens *et al.* (1981).

The most complete study of these measurements to date is that by Oort and Vonder Haar (1976), who investigated the extent to which they could employ radiation budget data from a number of satellites, together with measurements in the atmosphere and ocean from conventional sources, to arrive at a

Fig. 4.10 Components of the earth's radiation budget 1962-66 (from Vonder Haar and Suomi 1971).

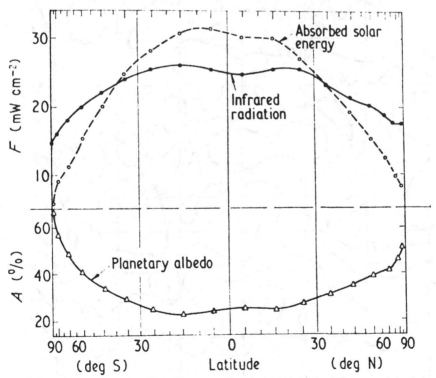

complete description of the energetics of the atmosphere–ocean system. Fig. 4.12 shows the components they considered; Fig. 4.13 illustrates what they did and shows some of their results. In Fig. 4.13 (*a*), F_{TA} was derived from the satellite data, and div T_A and S_A from conventional atmospheric data; F_{BA} could then be found since

$$F_{BA} = F_{TA} - \text{div } T_A - S_A \tag{4.1}$$

Referring now to Fig 4.13 (*b*), F_{BA} having been found from equation 4.1 above, S_0 being estimated from observations within the oceans, the remaining quantity div T_0 can be found from:

$$\text{div } T_0 = F_{BA} - S_0 \tag{4.2}$$

Fig. 4.11. Map of net radiation in Wcm^{-2} at the top of the atmosphere for August 1975 deduced from the Nimbus 6 ERB narrow-angle scanning channel observations (after Smith *et al.* 1977). Notice the large radiation deficit over the Sahara desert.

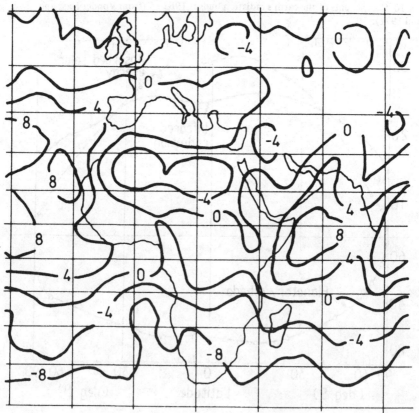

The further storage quantities S_L and S_I in Fig. 4.12 are very small by comparison and were neglected. The results of the error analysis carried out by Oort and Vonder Haar suggest that typical errors in their derivation of the quantities F_{TA} and div T_A were 10 W m^{-2}, in F_{BA} ~15 W m^{-2} and in div T_0 ~20 W m^{-2}. These errors can be compared with the mean solar input of 350 W m^{-2} and typical magnitudes of the above four quantities of 50 W m^{-2}. They therefore deduce that the pattern of oceanic heat transport which results from their study is probably not highly accurate but is nevertheless likely to be broadly correct. Some evidence for the correctness of their error estimates, so far as F_{TA} is concerned, comes from a study of the globally averaged net flux by Ellis *et al.* (1978) shown

Fig 4.12 Schematic of the different terms in the earth's energy budget (after Oort and Vonder Haar 1976). F_{TA} and F_{BA} are the net radiation fluxes at the top and bottom of the atmosphere respectively. S_A, S_I, S_L and S_O are the rates of storage energy in atmosphere, snow and ice, land and ocean. T_A and T_O are the horizontal transports of energy in the atmosphere and the ocean.

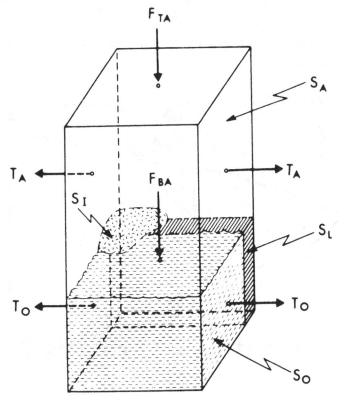

Fig. 4.13 Annual variation of various components of the heat budget of the northern hemisphere (after Oort and Vonder Haar 1976). See Fig. 4.12 for key to symbols.

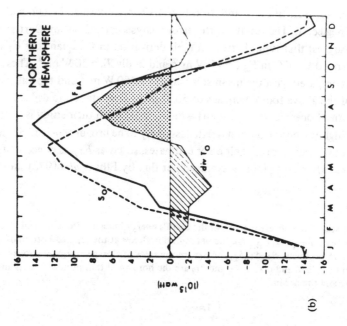

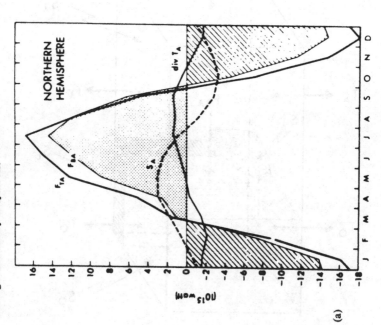

in Fig. 4.14. Consistency between the results from different sets of data was obtained; also the measured variation is consistent with completely independent measurements of variations in the heat storage of the oceans.

The work of Oort and Vonder Haar illustrates one particular aim of radiation budget studies, namely, the study of atmospheric energetics on the regional and global scales. In particular they demonstrate how the surface radiation budget F_{BA} and the oceanic heat transport div T_0 can be deduced. No other means is currently available for estimating the latter of these quantities. For climate research it is clearly necessary to know the size and scale of inter-annual variations in the various components describing the atmosphere's ener-getics. In particular, to assess the role of the oceans in climatic variation, we need to know to what extent the partitioning of heat transport between the atmosphere and the ocean varies from year to year. More accurate radiation budget measurements from satellites are therefore required. The probable error in F_{TA} when averaged over several days and over areas of, say, 1000 km square should be less than 5 W m^{-2}.

Another important objective of radiation budget measurements is the elucidation of the problem of cloud-radiation feedback. An understanding of this problem is central to the debate concerning how the average surface tem-perature might change as a result of factors such as changing solar constant or changing atmospheric carbon-dioxide content. To illustrate the problem simply, we can write for the average balance at the top of the atmosphere between the

Fig. 4.14 Annual variation of the globally integrated radiative flux at the top of the atmosphere as obtained from two separate data sets, the 29 month set being selected from Nimbus, ESSA and NOAA satellites for the period 1964-71 and the Nimbus 6 data set being from the ERB on Nimbus 6 for July 1975-June 1976 (after Ellis *et al.* 1978).

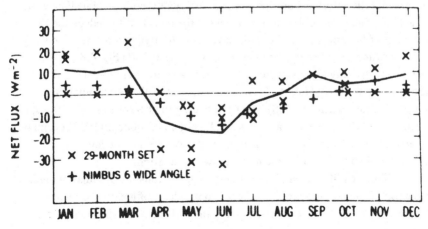

solar input $\frac{1}{4}S$ (S being the solar constant) and the outgoing reflected solar radiation (α is the albedo $\simeq 0.33$) and emitted radiation F

$$\frac{1}{4}S(1-\alpha) = F \tag{4.3}$$
$$= \epsilon \sigma T_s^{4}$$

where T_s is a mean surface temperature ($\simeq 280$ K), ϵ an emissivity of the earth-atmosphere system ($\simeq 0.7$) and σ is Stefan's constant. From equation (4.3) we find that for constant α and ϵ, i.e. for no cloud-radiation feedback, a 1% change in S leads to a 0.7 K change in T_s and a 0.01 change in α leads to a 1 K change in T_s. Assuming the mixed layer of the oceans which has a mean depth ~ 100 m, is involved in the changes as well as the atmosphere, the time constant of change is about 4 years.

To tackle the problem of cloud-radiation feedback, therefore, it is necessary to find out first how much the quantities α and F change from year to year and what is the scale of their variations in space and time, and then whether changes in the mean value of T_s can be associated with changes in the radiation budget parameters. Further, in order to arrive at an understanding of the cause of the changes, careful measurements of cloud amount and height (especially cloud top height), must be made associated with the radiation budget measurements. Since the size of the variations from year to year in the various parameters will probably be only a few per cent, we require continuous measurements over a number of years of the incoming solar radiation and the emitted long-wave radiation to better than 1 % accuracy, and measurements of albedo to better than 0.01. The accuracies quoted apply to averages over several days and over areas of $\sim 10^6$ km^2.

To what extent can these accuracies of measurement be achieved? There is certainly no problem in their achievement so far as sensor technology is concerned; satisfactory calibration to the accuracy quoted is possible if carefully carried out. The main problems arise from (1) the fact that, to deduce the flux at the top of the atmosphere from measurements at satellite altitude, modelling of the variation of radiance with angle has to be carried out (Fig. 4.8); and (2) the fact that any given satellite samples the radiation field preferentially at particular times of day (Fig. 4.15). In an assessment of the earth-radiation budget satellite system (ERBSS) being planned by the US for the early 1980s, in which a set of three satellites are involved, Woerner and Cooper (1977) estimate the errors arising from these two factors to be about 1% for the long-wave radiation and about 0.5% for the net short-wave radiation.

Thus it will be possible, with satellite measurements of radiation budget at the top of the atmosphere, to address two problems of great importance to climate research. These are the pattern and the variability of heat transport by

the oceans and the sign and magnitude of cloud-radiation feedback. If the
ERBSS system achieves sufficient accuracy, careful data analysis should result
in considerable progress on these two crucial problems by the end of the decade.

Fig. 4.15 Coverage in local time and latitude provided by three satellites, namely two
TIROS satellites in sun-synchronous orbit at 0800 and 1500 local time respectively to-
gether with an Atmospheric Explorer-type satellite in an orbit of 50° inclination (after
Woerner and Cooper 1977).

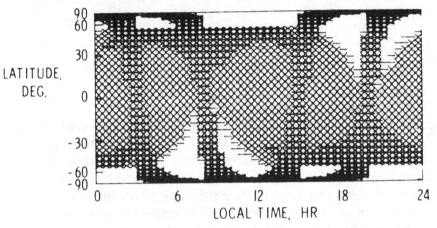

5

PRINCIPLES OF REMOTE TEMPERATURE SOUNDING

5.1   Introduction

At any frequency or wavelength in the infrared or microwave regions where an atmospheric constituent absorbs radiation, it will also emit thermal radiation according to Kirchhoff's Law. Thus the radiance leaving the top of the atmosphere will be a function of the distribution of the emitting gas and the distribution of temperature throughout the atmosphere. We would expect, therefore, that measurements of radiance should contain some information about both of these distributions. Measurements by satellite instruments using this principle have used two distinct kinds of viewing geometry: (1) the near-nadir view in which radiation is observed leaving the atmosphere in directions near to the local vertical, (2) the limb view (Fig. 5.1), in which radiation leaving the atmosphere nearly tangentially is observed. In the first of these, which has been

Fig. 5.1 Geometry of limb measurements illustrating the tangent height h and the projected thickness of an arbitrary layer at height z.

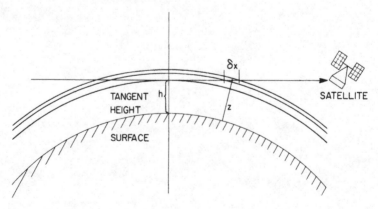

employed for most temperature sounding experiments, selection of the altitudes observed is carried out by the choice of appropriate spectral regions as explained below.

If we choose to look at radiation from gases whose distribution is well known, such as carbon dioxide or molecular oxygen, which are nearly uniformly mixed, then the radiance contains information about temperature only. Qualitatively, we can see that radiation emitted from low down in the atmosphere does not escape to space because it is reabsorbed at higher levels, whilst there is little emission from very high levels because of the decrease of density with height. As will be shown rigorously later, thermal emission comes from, and therefore has the characteristic temperature of, a fairly well-defined region of the atmosphere which is about 10–15 km thick, and whose altitude depends on the absorption coefficient. When the absorption coefficient is high, the emitting layer is high in the atmosphere, and when it is low, the emitting layer is low, so that by making measurements in several spectral regions we can sound a range of altitudes.

For the limb view, selection of altitudes to be observed is made by choosing the direction of view. The emission from a path of gas along a direction of view through the limb will be proportional to the emissivity of the gas and to the Planck function at some representative temperature such as that at the lowest point in the path. Thus if the concentration of the gas is known, the temperature may be deduced, and vice versa.

For a particular absorption band to be employed usefully for temperature sounding the following conditions need to be satisfied.

(1) The emitting constituent should be substantially uniformly mixed in the atmosphere so that the emitted radiation can be considered a function of the temperature distribution only. This is the case for molecular oxygen (which possesses an absorption band near 5 mm wavelength) up to about 100 km altitude, and also to within 1% or 2% for carbon dioxide (Bishof and Bolin 1966, Georgii and Jost 1969), probably up to about 100 km also (Hays and Olivero 1970). Absorption bands due to carbon dioxide occur near wavelengths of 15 μm and 4.3 μm.

(2) The absorption band involved should not be overlapped by bands of other atmospheric constituents. This is nearly true for the bands mentioned above, although some overlapping with bands of water-vapour, ozone and nitrous oxide occurs for which corrections have to be made.

(3) Local thermodynamic equilibrium (LTE) should apply, so that emission from the band is proportional to the Planck function. This requires that the population of the upper state belonging to the absorption band in question should be determined by the appropriate Boltzmann factor at

the local kinetic temperature, which will be the case if the probability of excitation by collision is considerably larger than the probability of de-excitation by radiation processes. As altitude increases, the collision frequency falls and at some altitude local thermodynamic equilibrium will no longer be a good approximation; for the 5 mm oxygen band the critical altitude will be well in the thermosphere (> 100 km); for the 15 μm band of CO_2 it is about 80 km and for the 4.3 μm CO_2 band about 35 km (Houghton 1969).

(4) The wavelength should be long enough that scattered solar radiation is insignificant compared with thermal emission. This is the case at wavelengths greater than about 4 or 5 μm.

The same requirements hold for composition sounding, except that the first item in the list is not applicable.

5.2   Theory of temperature sounding
5.2.1   *Nadir sounding*
We now consider, for a radiometer viewing radiation leaving the atmosphere near to the local vertical, the radiative transfer theory on which the idea of temperature sounding is based (Kaplan 1959, Houghton and Smith 1970). Consider first in an infinitely deep atmosphere which is horizontally stratified, a slice at temperature T containing a path length du of absorber in the vertical direction. At a frequency v where the absorption coefficient is k_v, under the assumption of LTE, Kirchhoff's law states that the emitted radiance from this slice in the vertical direction will be k_vd$uB_v(T)$ where $B_v(T)$ is the Planck function at frequency v and temperature T. Of this radiation a proportion τ_v will reach the top of the atmosphere where

$$\tau_v = \exp\left(-\int k_v \, du\right) \tag{5.1}$$

the integral being over the region of atmosphere between the slice and the top of the atmosphere. Integrating over all such slices to find the total radiance L_v at the top of the atmosphere

$$L_v = \int B_v(T) \, k_v \, du \, \exp\left(-\int k_v \, du\right)$$
$$= \int_0^1 B_v(T) \, d\tau_v \tag{5.2}$$

It is convenient to use as an altitude dependent variable $y = -\ln p$, where p is the pressure in atmospheres. In this case

$$L_v = \int B_v(T) \frac{d\tau_v}{dy} \, dy$$
$$= \int B_v(T) \, K(y) \, dy. \tag{5.3}$$

In other words L_ν is the weighted average of the black-body radiance, the weighting function $K(y)$ being $d\tau_\nu/dy$. The variable $y = -\ln p$ is employed instead of the height z as then the accompanying weighting function is more nearly independent of temperature.

For absorption lines in molecular bands in the lower atmosphere (< 50 km) collision broadening is dominant and Doppler broadening may, in general, be neglected. It is, therefore, instructive to work out the form of the weighting function $K(y)$ for a single frequency ν chosen in the wing of a collision broadened spectral line.

For a single collision broadened spectral line of strength s centred at frequency ν_0, the absorption coefficient k_ν for a path at pressure p atmospheres is

$$k_\nu = \frac{s\gamma_0 \, p\pi^{-1}}{(\nu-\nu_0)^2 + \gamma_0^2 \, p^2} \tag{5.4}$$

where γ_0 is its width at 1 atm pressure. In the line wing where $\nu-\nu_0 \gg \gamma_0 p$, k_ν is directly proportional to pressure. For a uniformly mixed absorber, du in equation (5.1) is proportional to dp so that $\tau_\nu = \exp(-\beta p^2)$ where β is a constant dependent on the line strength and width and the concentration of absorber. We then have

$$K(y) = 2 \left(\frac{p}{p_0} \right)^2 \left\{ \exp- \left(\frac{p}{p_0} \right)^2 \right\} \tag{5.5}$$

where p_0 is the pressure at which $K(y)$ is a maximum. This function is plotted in Fig. 5.2. The value of p_0, that is, the height of the maximum, depends on the particular wavelength which is chosen, but the half-width of the curve in units of $\ln p$ is independent of wavelength. Since for an isothermal atmosphere $\ln p$ is directly proportional to height z, the half-width in height units is approximately constant and equal to 10 km. The curve A of Fig. 5.2, therefore, represents the best that can be done as far as defining the region of origin of the radiation received at a given frequency. It is interesting to note that the effect of pressure broadening is to sharpen up the curve considerably; if Doppler broadening were dominant, the curve would have a considerably greater 'half-width'.

Since the width of lines, for instance in the 15 μm CO_2 band, varies from about 0.1 to 0.001 cm^{-1} over the range of atmospheric pressures involved, the bandwidth of a conventional spectroscopic system possessing the monochromatic response described above would have to be better than 0.1 cm^{-1} in the wings of a line and as high as 0.001 cm^{-1} near the line centres. Many of the radiometers designed for temperature sounding (Ch. 6) possess a spectral bandwidth which is very much larger than this – say, about 5 cm^{-1}. This is not only larger than the half-width of the lines but also than the mean spacing between the lines, so that radiation is received over a band in which the absorption varies very considerably. However, compared with the ideal monochromatic response which we have found above, the degradation in performance is not very large so far as emission from the lower atmosphere is concerned. This is because, for any spectral region containing several lines, most of the energy originates from the wings of the lines where the absorption coefficient varies relatively slowly with frequency. It is this radiation which is emitted by the lower atmosphere. The performance achieved with a resolution of a few wavenumbers can be illustrated by considering the weighting function appropriate to a spectral region for which

Fig. 5.2 Weighting functions for: A, a monochromatic frequency in the wing of a collision broadened line; B, a strong Elsasser band.

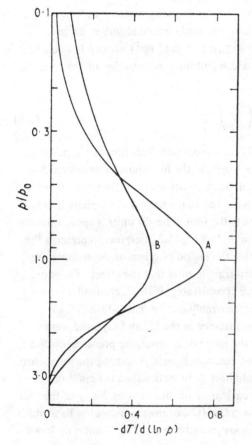

a strong Elsasser band model is appropriate. For such a region the mean transmission

$$\tau = 1 - \frac{2}{\pi^{1/2}} \int_0^{\beta' p} \exp(-x^2)\, dx = 1 - \text{erf}(\beta' p)$$

where β' depends on the local line strengths, widths and spacing. For this case

$$K(y) = \left(\frac{2}{\pi}\right)^{1/2} \frac{p}{p_0} \exp\left(-\frac{p^2}{2p_0{}^2}\right). \tag{5.6}$$

This weighting function is somewhat broader than that for the ideal monochromatic case giving somewhat poorer height resolution (see Fig. 5.2).

The above discussion has been based on a single frequency or a single narrow spectral interval for which the altitude of the peak of the weighting function depends on the absorption coefficient at the particular frequency and the concentration of emitting gas in the atmosphere. Because in any absorption band, the absorption coefficient varies rapidly with frequency, different frequencies will possess weighting functions peaking at different altitudes. Remote temperature sounding over a range of altitudes becomes possible if a set of frequency intervals (centred at frequencies $\nu_1, \ldots, \nu_i, \ldots, \nu_N$) can be chosen such that the corresponding intensities each originate from substantially different levels in the atmosphere and so that the total height range represented includes, if possible, all levels of interest.

In writing a general expression for the energy F_i incident on a detector in a remote sounding radiometer, we must include the emission from the lower boundary, the effect of the profile of the filter centred at ν_i and the other optical characteristics of the radiometer so that we have (Abel *et al.* 1970)

$$F_i = A\Omega \alpha_i(\Delta\nu_i) \left\{ B(\nu_i, T_0)\bar\tau(p_0) + \int_\infty^{-\ln p_0} B(\nu_i, T)\kappa_i(y)dy \right\} \tag{5.7}$$

with the weighting function

$$K_i(y) = \frac{\int_\nu f_{\nu_i}(d\tau_\nu/dy)\,d\nu}{\int f_{\nu_i}\,d\nu} \tag{5.8}$$

the mean transmission

$$\bar\tau(p) = \frac{\int_\nu f_{\nu_i}\,\tau_{\nu_i}(p)\,d\nu}{\int f_{\nu_i}\,d\nu} \tag{5.9}$$

and the effective band width

$$\Delta\nu_i = \int_\nu f_{\nu_i}\,d\nu \tag{5.10}$$

where A is the area of the entrance aperture;
            Ω is the solid angle of the field of view;

$B(\nu_i, T_0)$ is the Planck function at frequency ν_i corresponding to atmospheric temperature T at pressure p;

α_i is the frequency independent part of the transmission of the optics;

f_{ν_i} is the frequency dependent part of the optics transmission for the ith radiometer channel normalized to unit maximum;

p_0, T_0 are the pressure and temperature respectively at the lower boundary (ground or cloud).

The assumption has also been made that $B(\nu, T)$ is constant over the pass band of a radiometer channel so that it may be taken outside the integral over ν.

The variation with frequency or wavelength of $\bar{\tau}_\nu(p_0)$, the transmission of the atmosphere from the ground to space, is plotted in Fig. 5.3 both for the standard atmosphere and for each important constituent separately, showing the

Fig. 5.3 Absorption in the infrared for a vertical atmospheric path by a variety of constituents (from J.H. Shaw).

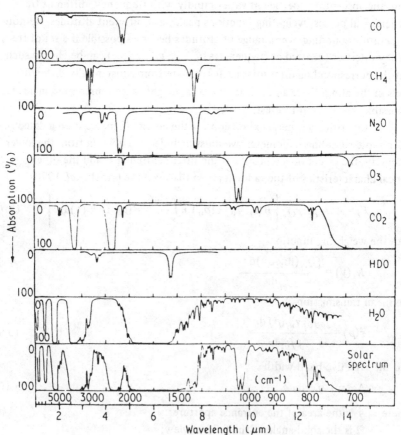

spectral region in which each contributes significant opacity to the whole. It should be clear from this that for a measurement in one of the atmospheric 'windows', for example at 11 μm, $\tau_\nu(p_0) \simeq 1$ and the first term in (5.7) will be large compared to the second, so that the measured brightness temperature is very nearly that of the ground. On the other hand, for a measurement in a narrow interval near 15 μm the transmission of the atmosphere is very low due to absorption by the strong ν_2 band of CO_2, and direct radiation from the ground cannot reach the satellite at all. The weighting function at this frequency will be large at high levels in the atmosphere. Fig. 5.3 should be compared with Fig. 5.4 which shows a recording of the emission spectrum of the earth plus atmosphere systems. In Fig. 5.4(*a*) radiation from the Q branch at the centre of the CO_2 band at 667 cm^{-1} originates at a temperature approaching 240 K, well within the stratosphere. Moving from there to higher frequencies the absorption coefficient falls and the brightness temperature falls to a value near 220 K, characteristic of the region near the tropopause, and then rises steadily as the peak of the appropriate weighting function falls in altitude, until a frequency of 800 cm^{-1} is reached when most of the radiation comes from the surface. By contrast, over the Antarctic (Fig. 5.4(*c*)) the temperature throughout the troposphere and stratosphere rises with altitude from around 185 K at the surface to 220 K in the stratosphere.

5.2.2 Limb sounding

In the case of limb sounding the equation of radiative transfer is essentially the same as for nadir sounding equation (5.2), but because of the geometry, the way in which it is used is rather different.

The radiance $L_\nu(h)$ at the spacecraft is found by integrating along the line of sight (Fig. 5.1) whose tangent altitude is h:

$$L_\nu(h) = \int_0^\infty B_\nu\left(T(x)\right) \frac{\mathrm{d}\tau_\nu(x)}{\mathrm{d}x} \, \mathrm{d}x \qquad (5.11)$$

where $T(x)$ is the temperature at distance x from the instrument, and $\tau_\nu(x)$ is the transmittance of the atmospheric path from x to the instrument.

A real instrument will have a field of view with a finite width in the vertical, so that the quantity measured is an integral of $L_\nu(h)$ over the field of view profile $f(h)$

$$L_\nu^{\mathrm{m}}(h) = \int L_\nu(h')f(h'-h)\mathrm{d}h' \qquad (5.12)$$

In the case of a vertical sounder it is generally possible to assume that the intensity is uniform over the field of view. This is not usually the case with a limb sounder, because of the rapid variation of radiance with the altitude of the tangent point.

Fig. 5.4 Examples of thermal emission spectra recorded by IRIS D on Nimbus 4. Radiances of black bodies at several temperatures are superimposed. (*a*) Sahara; (*b*) Mediterranean; (*c*) Antarctic; (all apodized) (from Hanel *et al.* 1971).

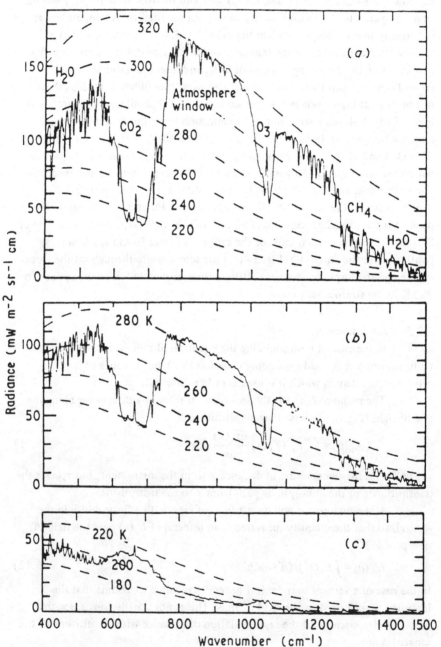

However, for the purpose of a simple-minded approximation we will ignore the effect of a finite field of view. We can see from Fig. 5.1 that a large fraction of the line of sight is at an altitude close to the tangent altitude h. If we make the approximation that all of the path is at the tangent point temperature, then we may write:

$$L_\nu(h) \simeq B_\nu(T(h)) \int_0^\infty \frac{d\tau_\nu(x)}{dx} \, dx$$
$$= B_\nu(T(h))(1-\tau_\nu(\infty))$$
$$= B_\nu(T(h)) \, \epsilon_\nu(h) \tag{5.13}$$

where $\epsilon_\nu(h)$ is the emissivity of the path. We see that if we can compute $\epsilon_\nu(h)$, as in the case of an absorber with known distribution, then we may derive the profile of Planck function $B_\nu(T(h))$, whilst if we know $T(h)$, we may derive $\epsilon_\nu(h)$ and hence the total absorber amount along the limb path.

The absorber amount in a limb path is clearly much greater than that in a vertical path, thus lending a major advantage to limb sounding over nadir sounding for the distribution of trace gases. For a uniformly mixed gas we may write

$$m = \int c \, \rho(x) \, dx$$

where c is the mass mixing ratio and $\rho(x)$ is the air density. We can relate x to z and h by

$$(R + h)^2 + x^2 = (R + z)^2$$

where R is the radius of the earth. Using $R \gg h, z$ we find

$$x^2 = 2R(z - h)$$

so that

$$m = \int c \, \rho(z) \sqrt{\frac{R}{2z}} \, dz \tag{5.14}$$

where we now express p as a function of height. Using the hydrostatic equation, and developing a similar expression for the absorber amount in a vertical path, it is easy to show that the tangent path contains about 75 times as much absorber as the vertical path.

In the case of temperature sounding, the accuracy to which we may find $B_\nu(T(h))$ depends on the size of ϵ_ν. The signal to noise ratio will be proportional to the emissivity of the path, thus setting an upper limit on the altitude that may be sounded. The lower limit to the altitude is determined by the breakdown of the approximation used above: if $\tau_\nu(\infty) \ll 1$ then the information is no longer coming from the tangent point, but rather from some point along the tangent path nearer to the spacecraft, and hence at a greater altitude. The range of altitudes which can be sounded will be determined by an inequality of the form

$$\epsilon_1 < \epsilon(h) < \epsilon_2$$

where ϵ_1 is some suitably small number and ϵ_2 is close to unity.

In the case of composition sounding, the accuracy to which we may find the absorber amount depends on the size of B_ν and on the derivative of ϵ_ν with respect to absorber amount. If we again use the strong Elsasser band model, we can write

$$L_\nu(h) = B_\nu(T(h)) \, \text{erf} \, (\beta''mp)$$

where β'' depends only on the spectral parameters of the absorption band, and p is pressure. The accuracy of a determination of m will be

$$\sigma(m) = \sigma(L) \bigg/ \left|\frac{\partial L}{\partial m}\right|$$

$$= \frac{\sigma(L)}{B_\nu} \left\{ \frac{2}{\pi^{1/2}} \beta''p \, \exp \left(-(\beta''pm)^2\right) \right\}^{-1} \tag{5.15}$$

which tends to zero for both $\beta''p$ small and for $\beta''pm$ large. There is only a finite range of altitudes in which good measurements can be made, coinciding approximately with the region in which the limb intensity is changing with altitude. The signal to noise ratio as a function of height is the same kind of shape as a vertical sounding weighting function.

The limb radiance is a linear functional of the profile of Planck function, so that the relationship between them can be expressed in terms of weighting functions, as in the case of vertical sounding. A typical set of weighting functions is shown in Fig. 5.5.

5.3   Calculation and measurement of weighting functions

In order to interpret remote sounding observations from a given instrument it is necessary to know the weighting functions $K_i(y)$ or limb emissivity for each channel. These calculations require knowledge of the optical characteristics of the radiometer and detailed spectral information regarding line positions, intensities, widths and shapes. Transmittances and hence weighting functions may be obtained by a numerical integration of an equation such as:

$$\tau(y, \infty) = \int d\nu \, f_{\nu_i} \exp \left\{ -\int_{z=y}^{\infty} dz \, \textstyle\sum_j k_j f_j(\nu, z) \, \rho(z) \right\} \tag{5.16}$$

where f_{ν_i} is the instrumental spectral response, k_j is the strength of the jth spectral line, $f_j(\nu, z)$ is its line shape factor and $\rho(z)$ is the absorber density. Efficient numerical evaluation of this function is a specialized task, for which the reader is referred to the literature.

Because of possible uncertainties in instrumental response, line strengths, shapes and positions, and in the numerical method, it is essential to verify the resulting transmittances against measurements. It is helpful if the measurements can be made in conditions which approximate the atmospheric situation as closely as possible. Any differences between measurements and calculations can be used to find the errors in the calculations, or in the assumptions about the spectral properties of the molecule. As a last resort, the measurements can be used to construct an empirical adjustment to the calculations.

In principle, the closest available correspondence between measurement and atmospheric weighting functions can be obtained by making measurements of transmittance of solar radiation through the atmosphere from a balloon platform. One such set of measurements was made by Batey & Abel (Houghton 1972) who used a radiometer similar to the Nimbus 5 SCR. A sample of their results is shown in Fig. 5.6. Agreement between atmospheric observations of

Fig. 5.5 A set of weighting functions for a limb sounder with an infinitesimal field of view (after Gille and House, 1971).

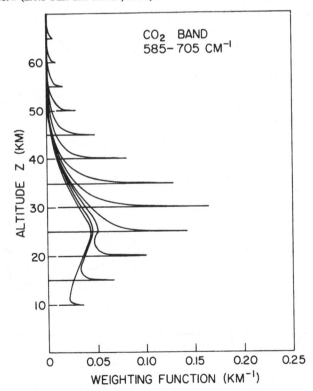

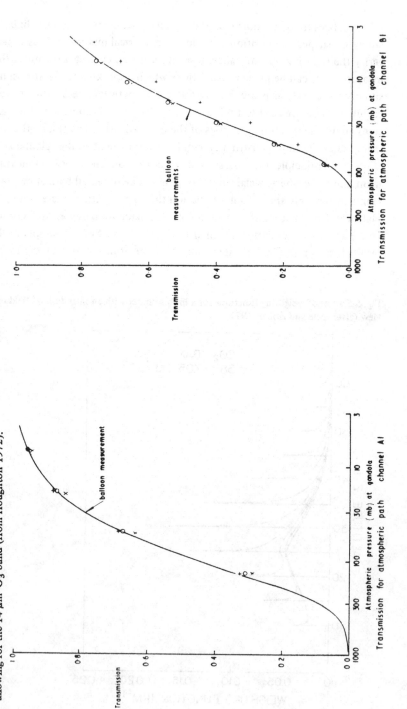

Fig. 5.6 Atmospheric transmission as measured (full line) at 15 μm wavelength by two channels (see 6.7 for a full description) of a radiometer similar to the Nimbus 5 SCR using the sun as a source with the radiometer mounted on an ascending balloon. Calculations from spectroscopic data were made for the atmospheric conditions (+). Laboratory transmission measurements adjusted theoretically for the atmospheric conditions were also made (0). The further points (x) denote the laboratory transmission measurements adjusted as above with an additional correction allowing for the 14 μm O_3 band (from Houghton 1972).

transmittance and its calculation by a combination of theory and laboratory measurement is reasonable although not within the 0.5% required. Significant improvement should occur when transmittances of laboratory paths over a range of temperature are available. Laboratory measurements can also be made. They have the disadvantage that the variable pressure path in the atmosphere cannot be simulated, but the advantage that path length, pressure, temperatures, etc., can be varied at will to support or disprove conjectures about the reason for disagreement between measurements and calculations. It is convenient to use absorber amounts and pressures that correspond approximately to those in the atmosphere. The Curtis–Godson (CG) approximation (Goody 1964) may be used for this purpose. For example, for carbon dioxide, the absorber amount between pressure level, p mbar, and space is approximately $5 \times 10^{-3} \, p$ g cm^{-2}, while the CG mean pressure is $p/2$. This situation can be obtained in a cell of length 500 cm and pressure $p/2$ mbar. The length of the cell is conveniently independent of the altitude simulated, so a run through the equivalent atmospheres can be carried out by varying pressure only. Fig. 5.7 shows a sketch of apparatus which has been used for this purpose. Fig. 5.8 shows some typical results, in this case for a pressure modulator radiometer of the type described below (§6.8). The measured points are compared to (a) a simplified calculation, which assumes

Fig. 5.7 Apparatus for weighting measurements. Radiation from the hot source S is chopped (C) and passed into the multiple-path absorption cell by mirror M_1, cylindrical light pipe L and window (KCl) W. Radiation is returned from the cell via a similar optical arrangement to the radiometer R. M_3 is for calibration purposes (from Abel *et al.* 1966).

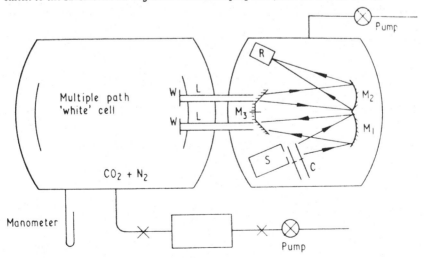

non-overlapping spectral lines, and (b) an exact calculation in which the transmission is obtained by integrating explicitly over all lines. Agreement between
the latter and the measurements is better than 1.5% everywhere.

Fig. 5.8 Transmittance of a 10m path of pure CO_2 gas as a function of pressure, as measured
by a pressure modulator radiometer with 8 mbar cell pressure and a cell length of 3 mm in
the apparatus of Fig. 5.7. Experiments points (·), simplified calculation (−), explicit
integration (×)

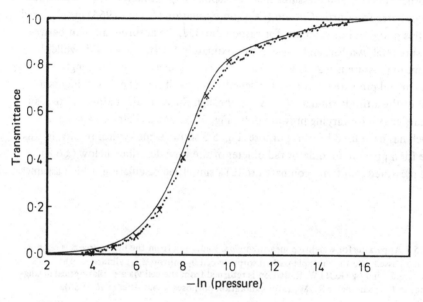

6

INSTRUMENTS FOR REMOTE TEMPERATURE SOUNDING

6.1 Introduction

In this chapter a brief description will be given of the satellite instruments which have been employed for the remote sounding of atmospheric temperature. Some individual temperature profiles deduced from measurements with them will also be presented, together with details of particular retrieval methods. A more general discussion of results will follow in Chs. 8 and 9 and a more general treatment of retrieval will be found in Ch. 7.

6.2 Principles of instrument design

The theory of remote temperature sounding has been described in Ch. 5 and some of the theoretical ideas on which the design of suitable instruments might be based are mentioned in §5.2. The important quantities which determine the quality of a given set of measurements are the vertical resolution or spread (§5.2) and the accuracy of measurement. To be adequate for meteorological purposes measurements of the mean temperature over layers ~ 200 mbar thick (~ 2 km near the surface, ~ 7 km at 12 km altitude) need to be made with a probable error of less than 1 K. The value of temperature measurements rapidly diminishes as the error becomes much greater than 1 K; for a given place, height and season the square root of the variance of temperature will normally be well under 10 K.

For radiance measurements in the $15 \, \mu m$ CO_2 band, an error of 1 K in temperature roughly corresponds to 1% in radiance. Because of the overlap of the weighting functions and because their vertical spread is in any case considerably greater than 200 mbar, individual radiance measurements must be made with considerably greater accuracy than this if the required accuracy is to be achieved.

Since meteorologists are more interested in spatial and temporal gradients of temperature than in absolute values, relative accuracy between similar instruments on the same spacecraft or between different channels of the same instrument is more important than absolute accuracy. The requirement for absolute accuracy is that there should be compatibility with the radiosonde network. An absolute accuracy of $\pm\frac{1}{2}$ % is adequate, therefore, realizing that the radiosonde network is probably no better than this. To realize compatibility between remote measurements and radiosondes, empirical adjustments can be made on the basis of detailed comparisons (Smith, Staelin & Houghton 1974). For relative accuracy, 0.1% is a useful aim. In §6.7 evidence is presented that this can be achieved.

The basic components of a radiometer for temperature sounding in a clear atmosphere are an optical system for defining the field of view and gathering the energy, a monochromator to define the spectral band pass and a detector. An instrument possessing a performance close to the theoretical limit would have a spectral resolution small compared with the width of an individual spectral line and an error of measurement equivalent to a temperature error of much less than 1 K, the time for a measurement for an instrument based on a satellite in fairly low polar orbit being much less than one second if adequate horizontal resolution is to be obtained. The number of independent spectral channels required to cover the lowest 50 km of a clear atmosphere is about 6 (*cf.* Ch. 7). However, because of the need to identify surface characteristics, clouds and water-vapour distribution as well, something more like twice this number of channels are required in practice.

A microwave radiometer (*cf.* §6.9) observing in the region of the 5 mm O_2 band can easily achieve the spectral resolution mentioned above but the precision of measurement we have mentioned demands at least a 0.1% measurement at this wavelength.

For radiometers operating in the infrared part of the spectrum the required precision of measurement is possible although not easy to achieve, but conventional spectroscopic instruments suitable for mounting on a satellite cannot approach a spectral resolution small compared with a line width. Different instruments can, however, be compared on the basis of their spectral resolution and their energy gathering power for a given size of instrument. For a dispersive instrument such as a prism or grating spectrometer or a Fabry-Perôt interferometer, the amount of radiation per unit spectral bandwidth depends simply upon the angular dispersion and the area of the dispersive element. As shown by the classic discussion of Jacquinot (1954) a grating passes significantly more energy for a given spectral resolution and size than a prism, and a Fabry-Perôt interferometer passes more than a grating by a factor of ~100 (see e.g. Houghton and Smith 1966. Ch.6, pp. 206-28). The Fabry-Perôt gains its advantage largely by employing circular apertures instead of the slit ones of prism or grating instruments. This circular symmetry is, of course, particularly appropriate when an extended source (such as the atmosphere below a satellite) is being observed. A similar energy-gathering power to the Fabry–Perôt interferometer may be obtained with a simple filter radiometer in which the spectral resolution is obtained by an interference filter. With such an instrument the smallest spectral bandwidth which can be achieved is limited currently to about $3 \, cm^{-1}$ at $667 \, cm^{-1}$ by the technical difficulties associated with manufacturing the filters. Such filters, since they employ materials of high refractive index, possess a light-gathering power up to an order of magnitude greater than that of an air-spaced Fabry–Perôt interferometer (Pidgeon and Smith 1962). For a set of spectral channels isolated by filters either a separate aperture and optical system is required for each channel or sequential measurements must be made through each filter in turn.

A further class of instrument is the Fourier transform interferometric spectrometer, an example of which is the Michelson interferometer. In this instrument, the whole spectrum is observed all of the time. As the path difference between the two beams within the instrument is changed, an interferogram results from which the spectrum is obtained by numerical transformation. This instrument combines the advantage of circular apertures with the 'multiplex advantage' which results from the simultaneous measurement of all the spectral elements. If there are N such elements there is a gain of sensitivity of $N^{1/2}$ compared to an instrument in which N elements are measured sequentially.

It was stated above that only of the order of twelve spectral channels are required. This means that the multiplex advantage of the interferometer is only of limited value for temperature sounding, although there are other applications requiring the examination of a wide range of spectra when the advantage possessed by the interferometer is very large indeed.

A number of instruments have been developed for flight on the Nimbus series of satellites which have followed one or other of the conventional designs mentioned above modified for space use. Other instruments have been developed using the principle of selective chopping which enables a very high effective spectral resolution to be achieved while retaining adequate energy gathering power. In these, paths of carbon dioxide itself are employed to filter the radiation emitted by carbon dioxide in the atmosphere – a technique which is similar in idea to that employed in infrared gas analysers. In the following sections instruments which have been flown or are being developed are described and some of the results already obtained with them presented.

6.3 Temperature sounding with MRIR

The first experiment to measure atmospheric temperature remotely from a satellite was mounted on TIROS 7. Measurements of the radiance of the atmosphere in the spectral interval 14.8-15.5 μm were made by including an appropriate filter in the MRIR radiometer which has already been described in §4.3. This single broad-band channel possesses a rather broad weighting function (Fig. 6.1) peaking at about 20 km altitude, so that the temperature observed was characteristic of the mean throughout the lower stratosphere. To obtain adequate accuracy with measurements from a spinning satellite, averages over several days were taken and global analyses made (Kennedy and Nordberg 1967).

6.4   Satellite infrared spectrometer (SIRS)

The satellite infrared spectrometer launched in April 1969 on Nimbus 3 was the first temperature sounding experiment with extended vertical coverage. Eight channels 5 cm^{-1} wide were located in the 15 μm ν_2 band of CO_2, corresponding to weighting functions which cover the atmosphere from the ground to around 35 km (Fig. 6.2).

SIRS is a grating spectrometer of Fastie-Ebert layout (Fig. 6.3) using a rotating mirror chopper. This is a highly reflecting rotating blade which regularly passes into the instrument field of view so that the spectrometer views the target and a reflected image of space with rapid alternation between the two (15 Hz), leading to a 15 Hz AC signal at the detector, a thermistor bolometer. Radiometric and wavelength calibrations are accomplished using an object mirror which moves so that from time to time the instrument views an internal black body of accurately known temperature with and without a broad-band calibration filter, the latter being to check for wavelength stability. Spectral selection is achieved by collimating the incoming radiation on to a fixed grating of 500 lines/cm and selecting the channels by placing eight slits on the image plane at the correct inclinations to the grating. The optics and electronics were designed to give a field of view of 12° (resolution at the surface about 200 km) and a dwell time of 6 s.

Smith *et al* (1970) have presented an extensive study of the accuracy of SIRS data compared with conventional radiosonde analyses and of the way in which it can be utilized in operational meteorological procedures. In this work they have employed a regression relation between the measured radiances and the temperature profile obtained by comparing SIRS radiance observations with carefully chosen radiosonde profiles nearly coincident in space and time. As was explained in §5.2, a need for accurate knowledge of weighting functions is thereby avoided. Their detailed retrieval method is described in Chs 7 and 8.

Figs. 6.4 and 6.5 show specimen temperature profiles retrieved in this way compared with radiosonde observations.

Fig. 6.1 Weighting functions for the 15 μm channel of the TIROS 7 MRIR (from Kennedy and Nordberg 1967).

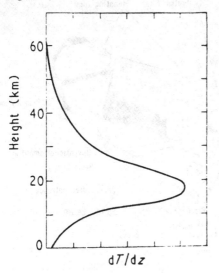

Fig. 6.2 Weighting functions for Nimbus 3 SIRS spectrometer (from Smith *et al.* 1970).

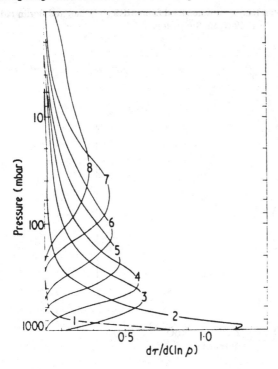

Fig. 6.3 The Nimbus 3 satellite infrared spectrometer (SIRS) (from Wark 1960).

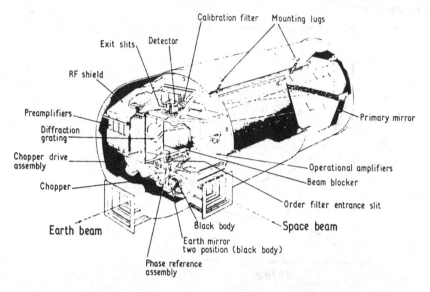

Fig. 6.4 Comparison between temperature retrieval from SIRS (full line) and Berlin radiosonde (broken line) on 7 June 1969 (from Smith *et al.* 1970).

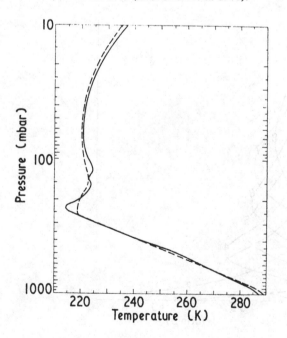

A further SIRS instrument was flown on Nimbus 4 (launched April 1970) which in addition to the channels for temperature sounding included a number for sounding water vapour.

Similar instrumentation for temperature sounding has also been built in the USSR. A scanning grating spectrometer covering the region 11–15 μm was launched on the Meteor 7 spacecraft in April 1971. The field of view, 15 km x 50 km, is smaller than that of the SIRS. Six channels have been employed for temperature retrievals by a similar method to that used for SIRS.

6.5   Infrared interferometer spectrometer (IRIS)

Nimbus 3 also carried a Michelson interferometer, IRIS, which measures the spectrum of the earth from 5 to 25 μm at a spectral resolution of 5 cm⁻¹. Nimbus 4 carried an improved version with still higher resolution, 2.8 cm⁻¹, and a version of IRIS flew on Mariner 9 to Mars in 1971 (Ch. 15).

Fig.6.6 shows the layout, with the multilayer beamsplitter on a potassium bromide substrate, the servo-controlled moving mirror drive, designed to advance the mirror evenly and without tilt, and the thermistor bolometer detector. A second complete interferometer is incorporated for reference pur-

Fig. 6.5 Comparison between temperature retrieval from SIRS (full line), and radiosonde (open circles) under partially cloudy conditions. The 'first guess' solution is also shown (broken line). (from Smith *et al.* 1970).

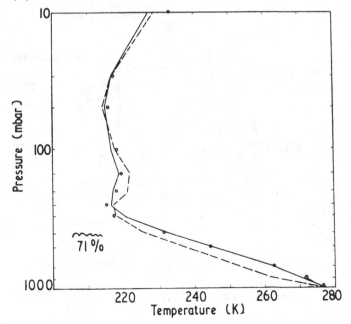

poses, using a neon discharge lamp as a source and a silicon photovoltaic cell as detector. The instrument views the earth via a gold-plated mirror which tilts to view space or an onboard black body when calibration is required.

The data are received at the ground as Fourier interferograms. After processing and Fourier transformation (Conrath *et al.* 1970, Hanel *et al.* 1970) spectra like those in Fig.5.4 are obtained. Information about the atmosphere's temperature profile can be obtained from the region of the spectrum around the $667\,cm^{-1}$ CO_2 band.

Information about the water-vapour and ozone distributions is also contained in the spectra and will be referred to elsewhere (Ch. 10).

A similar instrument has been mounted on the METEOR series of the USSR for temperature profile determination (Golovko and Spankuch 1978). Further exploitation of the interferometric technique for radiance measurements for atmospheric sounding has been proposed by Smith *et al.* (1979).

6.6   Filter radiometers

The first instrument for the remote sounding of atmospheric temperature, the MRIR on TIROS 7, which has already been mentioned (§4.3), was a simple filter instrument with one channel for temperature sounding. On later

Fig. 6.6 Schematic diagram of infrared interferometer spectrometer (IRIS) on Nimbus 3 (from Hanel *et al.* 1970).

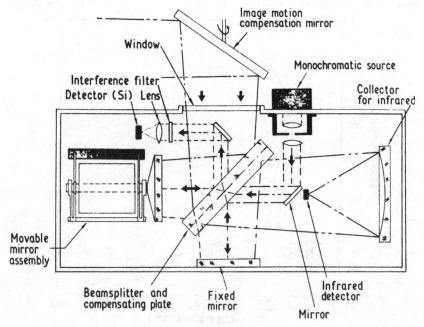

satellites a number of multichannel filter instruments has been flown, in particular the Infrared Temperature Profile Radiometer (ITPR) on Nimbus 5 (Smith *et al*. 1972), the High Resolution Infrared Radiation Sounder (HIRS) on Nimbus 6 (Smith *et al* 1975), and the Vertical Temperature Profile Radiometer (VTPR) (Schwalb 1972) on the TIROS operational satellites. The latest operational TIROS series (TIROS N) includes a HIRS for temperature sounding of the troposphere and lower stratosphere.

Other instruments which have included some filter channels for observations of temperature and composition are the Selective Chopper Radiometer (SCR) on Nimbus 4 and 5 (*cf.* §6.7) and various limb-sounding radiometers which are described in a separate section.

The ITPR on Nimbus 5 possessed seven channels, each with a separate optical system, four in the 15 μm CO_2 band for tropospheric temperature determination, one near 20 μm for water-vapour determination and two window channels at 11 μm and 3.8 μm respectively. A particular feature of the instrument was its narrow field of view (1.5° circular). Successive measurements could be made while scanning over a grid containing about 140 elements in a total time of 64 s. From the assumption that the temperature structure is uniform over the area of such a grid while the cloud structure may vary, and utilizing observations in both the window regions, a retrieval method could be developed which adequately eliminated the effect of complex cloud structure (Smith, Staelin and Houghton 1974).

The VTPR which has flown on the operational satellites NOAA 2, 3, 4 and 5, is an eight-channel instrument with a field of view of 2.235° x 2.235° scanning across the orbital track.

The HIRS on Nimbus 6 and on the TIROS-N operational series is an elaborate instrument with up to twenty channels in both the 15 μm and the 4.3 μm bands (Table 6.1). The advantages of including 4.3 μm band channels are (1) because of the much higher dependence of radiance on temperature at 4.3 μm, the temperature error introduced by broken cold clouds obstructing radiation from warmer regions below is considerably less than at 15 μm; (2) because the derivative dB/dT (where B is the Planck function at temperature T) falls off with altitude in the troposphere with a much larger gradient at 4.3 μm than at 15 μm, the effective weighting functions for temperature (i.e. $d\tau/d(-\log p)$ in equation 5.3 multiplied by $dB/d\bar{T}$, $\bar{T}$ being a mean temperature profile) are much sharper at 4.3 μm than at 15 μm. Figs. 6.7 and 6.8 illustrate the sharper effective weighting functions in the lower troposphere for the 4.3 μm band.

The optical system of the HIRS is illustrated in Fig. 6.9. Common entrance optics are followed by a dichroic element which divides the incident radiation into the long-wave channels (1-10) which are transmitted and the short-

Table 6.1 *Functions of the HIRS channels.*

Channel number	Channel central wave-number	Central wavelength (μm)	Principal absorbing constituents	Level of peak energy contribution	Purpose of the radiance observation
1	668	15.0	CO_2	30 mb	Temperature sounding. The 15 μm band channels provide better sensitivity to the temperature of relatively cold regions of the atmosphere than can be achieved with the 4.3 μm band channels. Radiances in Channels 5, 6, and 7 are also used to calculate the heights and amounts of cloud within the HIRS field of view.
2	680	14.7	CO_2	60 mb	
3	690	14.4	CO_2	100 mb	
4	703	14.2	CO_2	250 mb	
5	716	14.0	CO_2	500 mb	
6	733	13.6	CO_2/H_2O	750 mb	
7	749	13.4	CO_2/H_2O	900 mb	
8	900	11.0	Window	Surface	Surface temperature and cloud detection.
9	1030	9.7	Window	Surface	
10	1225	8.2	H_2O	900 mb	Water vapour sounding. Provide water vapour corrections for CO_2 and window channels. The 6.7 μm channel is also used to detect thin cirrus cloud.
11	1365	7.3	H_2O	600 mb	
12	1488	6.7	H_2O	400 mb	
13	2190	4.57	N_2O	950 mb	Temperature sounding. The 4.3 μm band channels provide better sensitivity to the temperature of relatively warm regions of the
14	2210	4.52	N_2O	850 mb	
15	2240	4.46	CO_2/N_2O	700 mb	
16	2270	4.40	CO_2/N_2O	600 mb	

17	2360	4.24	CO$_2$	5 mb	atmosphere than can be achieved with the 15 μm band channels. Also, the short-wavelength radiances are less sensitive to clouds than those for the 15 μm region.
18	2515	3.98	Window	Surface	Surface temperature. Much less sensitive to clouds and H$_2$O than 11 μm window. Used with 11 μm channel to detect cloud contamination and derive surface temperature under partly cloudy sky conditions.
19	2660	3.76	Window	Surface	
20	14 500	0.69	Window	Cloud	Cloud detection. Used during the day with window channels to define clear fields of view and to specify any reflected solar contributions to the 3.7 μm channel.

Fig. 6.7 Weighting functions $d\tau/d(-\log p)$ multiplied by dB/dT for the HIRS temperature and water-vapour channels (from Smith *et al.*1975).

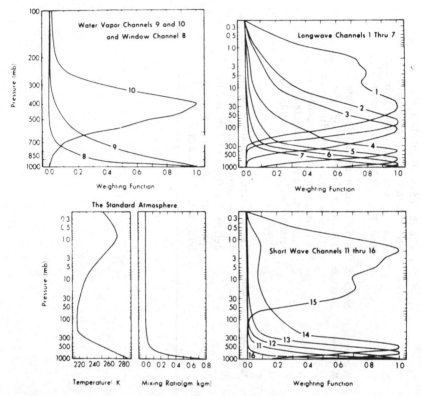

Fig. 6.8 Weighting functions (gradient of transmission with respect to log pressure) for instruments sounding the temperature of the lower atmosphere on the Nimbus 6 satellite: (*a*) 15 μm channels of HIRS. (*b*) 4.3 μm channels of HIRS, (*c*) channels of SCAMS (c.f. § 6.9; from Smith and Woolf 1976 and Staelin *et al.* 1975).

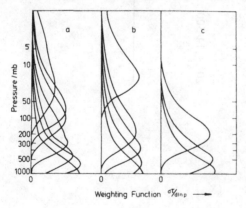

wave visible channels which are reflected. There are, therefore, two field stops for the system. Immediately behind the field stops, individual filters are arranged on a single filter wheel. Detectors cooled to ~ 120 K with a passive radiator are used for all infrared channels because of their high detectivity and short response time. The field of view of the instrument is circular, about $1.25°$ in angular width. The instrument scans in a similar way to ITPR. Details of the scan pattern are shown in Fig. 6.10.

The GOES 4 geostationary satellite launched in 1980 carried the VISSR (Visible Infrared Spin Scan Radiometer) Atmospheric Sounder, also known as

Fig. 6.9 Schematic of HIRS optical system (from Smith *et al.* 1975).

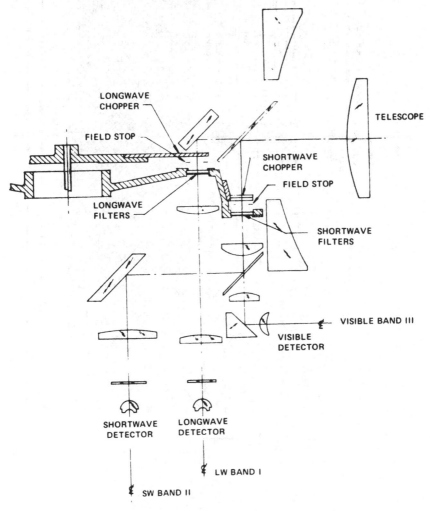

92

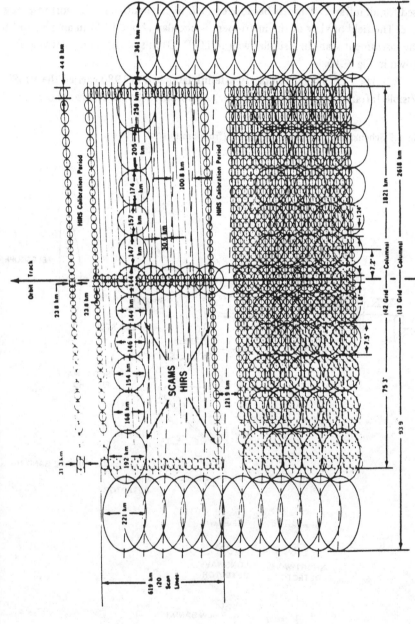

Fig. 6.10 HIRS and SCAMS scan patterns projected on Earth. (after Smith *et al.*, 1975).

the VAS, which combines visible channels for imaging cloud patterns with twelve infrared channels for temperature and water vapour sounding (Table 6.2). The satellite spins at 100 r.p.m. and achieves spatial coverage at resolutions of 1 km in the visible and 7 or 14 km in the infrared (depending on the channel). The VAS can be operated in two different modes: 1) a multi-spectral imaging (MSI) mode; 2) a dual sounding (DS) mode, for which the filter and scan mirror positions remain fixed during multiple spins of the spacecraft. Within each mode of operation there is a wide range of options regarding spatial resolution (7 or 14 km) spectral channels, spatial coverage and time frequency of observation. The purpose of the DS mode is to provide sufficient time of observation to achieve the required signal to noise ratio in the infrared temperature sounding channels. Although the spatial resolution which can be achieved from geostationary orbit is not as great as from a polar orbiting spacecraft, the great advantage of VAS is that it enables continuous observation of a wide region of the atmosphere. This is especially valuable in the study of the development of mesoscale systems (Smith *et al* 1982).

6.7   The selective chopper radiometer (SCR)

An infrared emission band generally contains many lines of width (defined as the frequency difference between points where the absorption co-efficient is half that at the centre) much less than 0.1 cm^{-1} under atmospheric conditions. Conventional spectroscopy is unable to resolve details of the structure

Table 6.2 *VAS instrument characteristics (from Smith et al. 1981)*

Spectral Channel	Central Wavelength (μm)	Weighting function peak (mb)	Absorbing Constituent
1	14.71	40	CO_2
2	14.45	70	CO_2
3	14.23	150	CO_2
4	13.99	450	CO_2
5	13.31	950	CO_2
6	4.52	850	CO_2
7	12.66	surface	H_2O
8	11.24	surface	window
9	7.25	600	H_2O
10	6.73	400	H_2O
11	4.44	500	CO_2
12	3.94	surface	window

near the line centres and, therefore, to select radiation from spectral regions where the absorption coefficient is very high. Work in the UK has been carried out jointly by teams at the Universities of Oxford and Reading (the Reading team transferred to Heriot–Watt University in 1970) and has been directed at using less conventional means to achieve an effective spectral resolution comparable with the width of the lines themselves, while collecting sufficient energy to make measurements of adequate accuracy.

It was first proposed (Houghton 1961, Smith 1961) to employ a Fabry–Perôt interferometer, the spacing between orders of which was matched to the rather uniform spacing between the lines in the ν_2 band of CO_2. This approach was superseded by the proposal (Smith and Pidgeon 1964) to employ a cell containing carbon dioxide itself to filter the radiation, which enables radiation

Fig. 6.11 Demonstrating the effect of selective absorption, weighting functions for: curve A, a 5 cm^{-1} wide interval near 690cm^{-1}; curve B, the same interval as for curve A but including a path of CO_2; curve C, an ideal weighting function for a monochromatic frequency in the wing of a spectral line (from Abel *et al.* 1970).

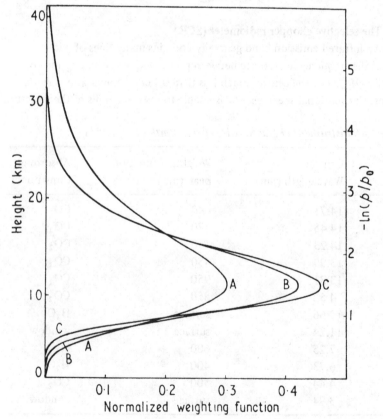

to be selected from regions where the absorption coefficient has a given value in the vicinity of many lines in the band (Peckham *et al.* 1967). The spectral resolution of such a system is very high, of the same order as the width of the lines in the absorbing CO_2 cell, that is typically 0.01 cm^{-1}, while, because radiation is collected from many lines, the total bandwidth and hence the energy grasp remain acceptable.

The selective chopper radiometer on Nimbus 4 (Houghton and Smith 1970, Abel *et al* 1970) consists of six independent filter radiometers, the spectral band pass of the interference filters varying between 3.5 cm^{-1} and about 10 cm^{-1}

For the four channels observing the lower levels of the atmosphere (<30 km) absorbing paths of CO_2 are introduced into the optical systems so that these channels are no longer sensitive to radiation originating near the centres of the absorption lines and, therefore, from regions of high altitude. The effective spectral resolution is, therefore, increased to the point where a weighting function almost as fine as the ideal case is obtained, but with the retention of a realistic energy grasp (Fig. 6.11).

Fig. 6.12 Demonstrating the effect of a selective chopping. Weighting function for: curve A, a 5cm^{-1} interval centred at 668cm^{-1}; curve B, the same interval as curve A but with a path of CO_2 of 1 cm at 0.05 atm pressure; curve C, (= curve A - curve B) the weighting function for selective chopping between conditions A and B (from Abel *et al.* 1970).

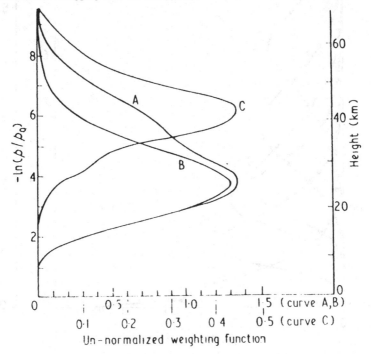

Weighting functions peaked at high levels are obtained by optically chopping the incoming radiation between two cells, one of which contains CO_2, the other being empty. An alternating signal is obtained which is dependent only on the radiation absorbed by the CO_2 in the absorbing cell, that is, the effective spectral response of the system is just those frequencies absorbed by the CO_2. The resulting weighting function can be made to peak well in the upper stratosphere (Fig. 6.12). Figure 6.13 shows the complete set of weighting functions used by the Nimbus 4 instrument, and Fig. 6.14 the optical design. The view through each cell is switched between Earth and space by a vibrating mirror, so that an AC signal appears at the detector and is amplified by phase-sensitive

Fig. 6.13 Set of weighting functions for SCR on Nimbus 4. Curves *A-F* refer to the six channels (from Abel *et al.* 1970).

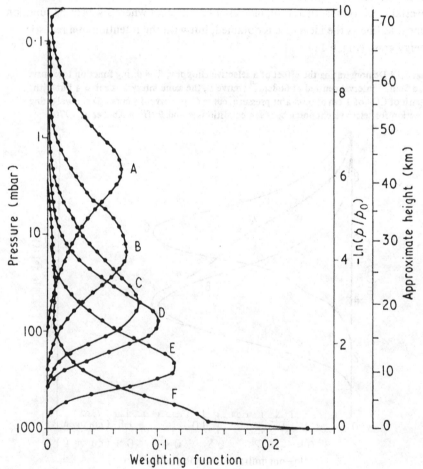

electronics tuned to the chopper frequency. The detector is a thermistor bolometer, as in SIRS and IRIS, and again calibration is achieved by the use of a
stepped object mirror which views the earth, space and a warm black body of
known temperature, in turn. With the SCR good temperature retrievals are
possible up to 45 or 50 km altitude, 15 km or so higher than with conventional
techniques. The structure of the stratosphere at these altitudes has been studied
on a global scale for the first time and a lot of new information about stratospheric warmings obtained (Figs. 6.15 and 6.16). Publication of two years of
Nimbus 4 data has been carried out by Oxford University Department of
Atmospheric Physics (1972).

For the Nimbus 5 satellite launched in December 1972 a more elaborate
SCR was built which possessed sixteen channels instead of six. The optical
system is shown in Fig. 6.17. A single 12.5 cm diameter aperture is common to
all the channels. The aperture is first divided into two by a rotating reflecting
chopper; both of the resulting beams are again divided into two by means of
dichroic beamsplitters which transmit one part of the spectrum while reflecting
another part. Four detectors are employed; in front of each detector a filter

Fig 6.14 Optical system of selective chopper radiometer (SCR) on Nimbus 4, channels
1 and 2. Inset shows single-cell radiometer (channels 3,4, 5 and 6) (from Abel *et al.* 1970).

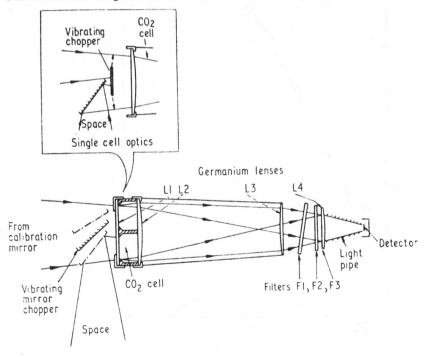

wheel positions four different filters of CO_2 cells in turn, making sixteen channels in all. These are listed in Table 6.3 together with a brief statement of the purpose of each. Improvements from the Nimbus 4 SCR are as follows.

(1) Because pyroelectric detectors having a better performance than thermistor bolometers were available, it was possible to have a much narrower field of view of $1\frac{1}{2}°$ corresponding to a horizontal field about 25 km diameter at the earth's surface compared with $\sim 10°$ for the Nimbus 4 SCR. This considerably increases the chance of a clear field

Fig 6.15 Comparison of retrieved temperature profile from SCR on Nimbus 4 (full line) with radiosonde and rocket sonde (broken line) from Wallops Island on 27 August 1970. CH is the cloud height and CA the cloud amount deducted from the retrieval procedure (from Barnett *et al.* 1972).

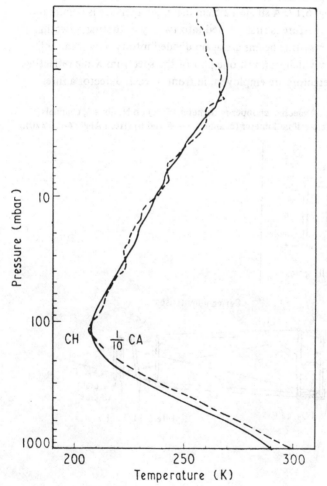

of view by making it possible to look through 'holes' in clouds and, in any case, makes it much more likely that cloud cover is uniform within the field.

(2) Four CO_2 channels instead of three sound the troposphere, and a channel at $520\,cm^{-1}$ for sounding water vapour was also included as in the Nimbus 4 SIRS instrument.

(3) Four channels for sounding for cirrus cloud were included. The presence of high cloud is much more of a handicap to obtaining good temperature retrievals than is low cloud (Hayden 1971) so that some specific information about high cloud as well as being useful in its own right enables better retrievals to be obtained (*cf.* §11.1).

(4) In addition to a window channel near $900\,cm^{-1}$, two window channels at shorter wavelengths were included which can be utilized at night (solar radiation is a problem during the day). They are at $3.5\,\mu m$ and

Fig. 6.16 Equivalent temperature (in K) observed by channel A of the Nimbus 4 SCR whose weighting function (Fig. 6.13) is centred near 45 km altitude for 4 January 1971. Note the very intense planetary wave of 'wave number one' (after Houghton 1972).

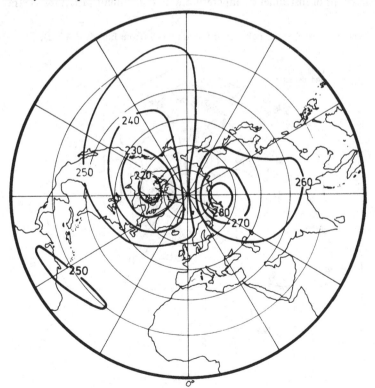

2.3 μm. Because the variation of Planck function with temperature is very different at the three wavelengths, it is possible from measurements at all three to build up a model of low cloud cover as well as enabling the surface temperature to be determined except when the cloud cover is complete.

Ten years after launch the Nimbus 5 SCR was still functioning in most of its channels.

Calibration of remote-sounding infrared instruments is carried out by pointing them in turn at a radiation zero and at a black body at a known temperature. Cold space provides an excellent radiation zero; black bodies of >99.5% emissivity can be constructed, and the blackbody temperature can easily be measured to better than 0.25 K. A linearity of detector response with respect to incident radiation of better than 0.25% can also be achieved. Care must also be taken that the optical system for calibration is sufficiently identical to that for measurement. With careful design, therefore, an accuracy of an absolute measurement of $\sim\frac{1}{2}$% can be achieved – a requirement mentioned in §6.2.

For a consideration of relative accuracy relevant parameters are detector noise, variations of instrument temperature with consequent variations in stray

Fig. 6.17 The selective chopper radiometer for Nimbus 5 (from Ellis *et al.* 1972).

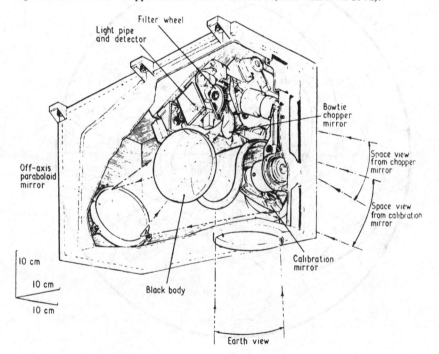

Table 6.3 *Nimbus 5 SCR channels.*

Channel number	Filter centre (cm^{-1})	Integrated equivalent width (cm^{-1})	CO_2 absorption path (cm)	CO_2 pressure (atm)	Equivalent width after selective absorption (cm^{-1})	Purpose
A1	668	10				Temperature sounding 0–18 km: CO_2 R branch
A2	689	10				
A3	707	10				
A4	726	15				
B1	668	6	0.3	0	6.0	'Selective chopping' channels for temperature sounding 25–45 km: CO_2 Q branch
B2	668	6	0.3	0.03	5.4	
B3	668	6	0.3	0.1	4.8	
B4	668	6	0.3	0.3	3.7	
C1	<110†	–				Cirrus detection
C2	202	25				H_2O emission
C3	536	15				Window channel/day and night
C4	859	100				
D1	3710	80				Reflected sunlight from cirrus
D2	3805	100				
D3	4550	400				Window channels: night-time only
D4	2817	60				

† Edge filter at 110 cm^{-1}.

radiation, detector response, electronic gain, etc. An upper limit to the overall random errors of a typical instrument can be obtained by looking at a location where small atmospheric variations are expected, such as the upper troposphere and lower stratosphere over the equator. Such a comparison has been carried out with the Nimbus 5 SCR. Table 6.4, which is taken from Barnett *et al.* (1975), lists the standard deviation of daily values of radiance for different channels of the Nimbus 5 selective chopper radiometer over 0°E at the equator and over Berlin. For comparison we might note that standard deviations of temperature as measured by the radiosonde over Berlin are 1.5 K at 50 mbar and 1.9 K at 5 mbar (Hawson 1970). From these figures it is seen that most of the variability over Berlin and some of that over the equator come from atmospheric variations. The residual due to instrumental variations is certainly less than 0.1 mW m^{-2} (cm^{-1})$^{-1}$ sr^{-1} (about 0.1 K in equivalent temperature) for channels B4 and A1 and cannot be much greater than this value for the selectively chopped channels B12, B23 and B34. The aim of 0.1% mentioned in §6.2 has, therefore, been achieved.

6.8   The pressure-modulator radiometer (PMR)

If a selective chopper radiometer (§6.7) is required to select radiation from very close to the centres of the lines, that is, emitted from as high an altitude as possible, problems of optical balancing of the double beam system become acute. A technique to overcome this problem employs a single cell of gas in which the pressure is modulated. The transmission of the cell and hence the radiation falling on the detector is modulated only at frequencies which lie within the absorption lines of the gas. Additional filtering requirements are not severe, as all that is required is to isolate the particular absorption band of the gas in question. Any emission from the gas itself is eliminated by the calibration procedure. An instrument employing the pressure modulator technique can be

Table 6.4 *Standard deviation of Nimbus-5 SCR radiance during June 1973*

Channel designation	Approx. pressure at peak of contribution function, mbar	0°N 0°E	52°N 10°E
B12	1.5	1.5	1.7
B23	4	1.2	1.7
B34	7	0.69	1.0
B4	50	0.08	0.46
A1	60	0.13	1.0
A2D	100	0.20	0.87

Units are mW m^{-2} (cm^{-1})$^{-1}$ sr^{-1}.

simple and robust and is almost ideal in the way it is sensitive only to energy at the required frequencies, while rejecting the rest. Sensitivity is limited only by the performance of the detector. A balloon flight of an instrument using this technique was carried out in 1970 (Taylor *et al.* 1972).

A PMR was mounted on the Nimbus 6 satellite launched in 1976 (Curtis *et al.* 1974). Fig. 6.18 shows a diagram of a pressure modulator cell. For temperature sounding the atmosphere above 50 km a pressure of a few millibars in the cell is required. The piston on its springs has a natural resonance frequency of about 15 Hz, at which frequency leakage of gas past the piston is not a problem.

Fig. 6.19 illustrates how the technique operates for a typical spectral line in the 15 μm CO_2 band. The top set of curves in Fig. 6.19(*a*) shows the transmission in a vertical path of atmosphere from various levels to the top of the atmosphere. Fig. 6.19(*b*) shows the effective transmission of a PMR cell when sufficient gas is present for the lines to be strongly absorbing at the centres, and Fig. 6.19(*c*) shows the effective transmission when there is weak absorption near

Fig. 6.18 Pressure modulator cell for PMR on Nimbus 6 (from Curtis *et al.* 1974).

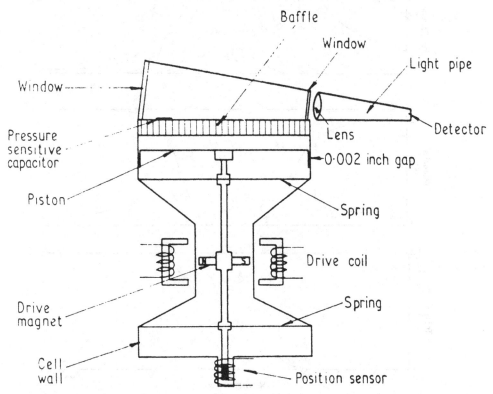

the line centres. Taylor *et al* (1972) show that the weighting function in the case shown in Fig. 6.19(*b*) is given by

$$\kappa(y) = \frac{p^{*2}}{(1 + p^{*2})^{3/2}}$$

where $y = -\ln p$ (p being the pressure in atmospheres) and $p^* = ap/p_{co}$, p_{co} being

Fig. 6.19 Illustrating the technique of pressure modulation by various plots near the centre of a single line in the ν_2 CO_2 band of strength 2cm^{-1} (atm cm)$^{-1}$. (*a*) 1 minus transmission plotted against wavenumber for various vertical atmospheric paths from the level of pressure p atmospheres to top of atmosphere. The number of each curve is the value of ln p. (*b*) Transmission for PMR cell 6 cm long at pressure 0.7 mbar (curve *A*), 2 mbar (curve *B*). Curve *C* (=curve *A*-curve *B*) is the effective transmission of the radiometer. (*c*) Transmission for PMR cell 1 cm long and PMR cell pressures 0.25 mbar (curve *A*), and 1.0 mbar (curve *B*). Curve *C* (=curve *A* - curve *B*) is the effective transmission of the radiometer. Curve *D* is curve *C* Doppler shifted by the amount which would occur for a Nimbus satellite if the radiometer viewed at 10° to the vertical (from Curtis *et al.* 1974).

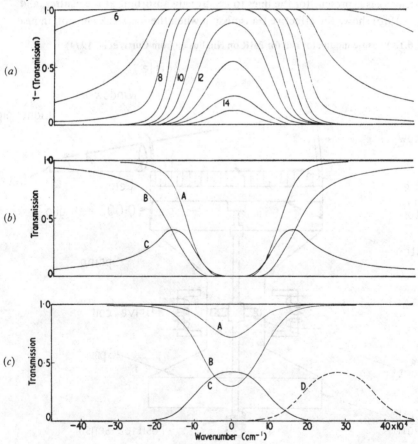

the mean pressure of gas in the cell and a a constant depending on the length of the cell compared with the concentration of CO_2 in the atmosphere. Fig. 6.20 shows weighting functions for different conditions integrated over all the lines in the band demonstrating that by varying the pressure of gas in the cell, the temperature may be monitored over the range of altitudes from 45 km to nearly 90 km.

A further way in which scanning in altitude of the emitting region can be devised is by utilizing the Doppler shift which arises between the atmospheric emission lines and the absorption lines of the gas in the cell when there is relative motion along the line of sight between the radiometer and the emitting atmosphere. Since the speed of a Nimbus satellite is about twenty times molecular speeds at normal atmospheric temperatures, only about 5% of the satellite's velocity is required to produce a Doppler shift equal to the Doppler line width. By varying the Doppler shift it is possible to scan the absorption lines across the atmospheric emission lines (Fig. 6.19(c)). The Nimbus 6 PMR instrument was designed so that the direction of view could be altered from vertically downwards to 15° from the nadir along the direction of flight, thus introducing varying Doppler shifts. By this means the optical depth (and hence the altitude) being probed could be varied (Fig. 6.21).

Fig. 6.20 Weighting functions for a pressure modulator radiometer having a cell 6cm long operating in the ν_2 band of CO_2 with different cell pressures. The number of the curves is mean cell pressure in mbar (from Curtis *et al.* 1974).

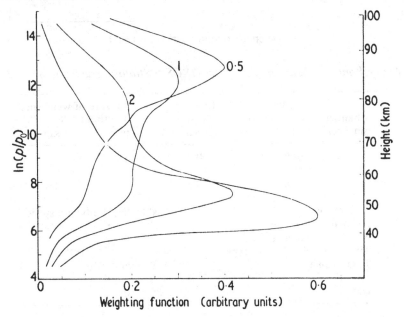

A set of three pressure modulator radiometers form the stratospheric sounding unit of the TIROS N operational sounder (Table 6.5). The modulator cells, each 1 cm long, contain CO_2 at mean pressures of 100, 35 and 10 mbar

Fig. 6.21 Weighting function for a pressure modulator radiometer housing a cell 1 cm long containing CO_2 at a mean pressure of 0.5 mbar when viewing at different angles to the nadir, hence allowing scanning of the emitting lines from the atmosphere across the absorbing lines in the cell by the Doppler shift due to the relative motion between atmosphere and instrument. The number of each curve is the angle to the nadir in degrees (from Curtis *et al.* 1974).

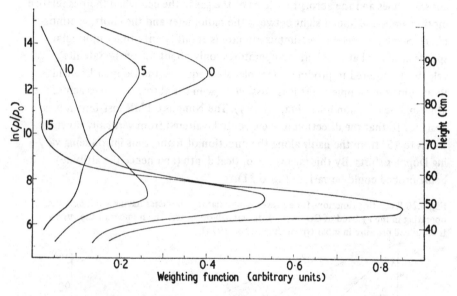

Table 6.5. *Principal characteristics of the TIROS-N Stratospheric Sounding Unit*

Channel number	Central wavenumber (cm^{-1})	Cell Pressure (mbar)	Pressure of weighting function peak	
			mbar	km
1	668	100	15	29
2	668	35	5	37
3	668	10	1.5	45

Calibration	Stable blackbody and space
Angular field of view	10°
Number of earth views/line	8
Time interval between steps	4 s
Total scan angle	±40° from nadir
Scan time	32 s
Data rate	480 bits/s

respectively. The weighting functions appropriate to the three channels are shown in Fig. 6.22. A fuller discussion of the design and performance of PMRs is given by Taylor (1983).

Fig. 6.22 Weighting functions for those channels used for stratospheric analysis. The labels indicate the channel designations within the TIROS operational vertical sounder. The associated instruments are: stratospheric sounding unit (solid line); microwave sounding unit (dotted line); high-resolution infrared sounder (dashed line) (from Miller *et al.* 1980).

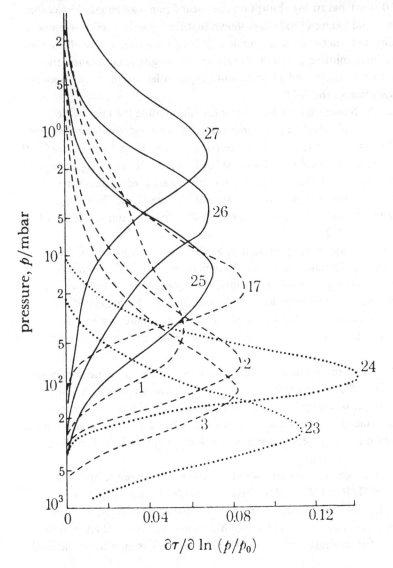

6.9 Microwave radiometers

All the instruments mentioned so far observe infrared emission from the infrared CO_2 bands. The most severe problem when sounding tropospheric temperature with such instruments is the presence of clouds.

Cloud which does not contain large droplets or large ice crystals is substantially transparent to radiation at microwave frequencies. Hence, temperature sounding at these long wavelengths has an important advantage over infrared sounding. Oxygen, which is uniformly distributed to high levels in the atmosphere (~ 100 km), has strong absorption lines near 5 mm wavelength. Meeks and Lilley (1963) and Lenoir (1968) have shown that these can be used in a temperature sounding experiment up to a considerable height although above 50 km the effect of Zeeman splitting of the lines in the earth's magnetic field causes the absorption to become dependent on azimuth angle, polarization and the geographical variation of the field.

On the Nimbus E (Nimbus 5 after launch) satellite the microwave sounding instrument called NEMS (Nimbus E microwave spectrometer) had five channels (Staelin *et al* 1972), one in a water-vapour absorption band (22.24 GHz), one in an atmospheric window (31.40 GHz) and three in the 5 mm O_2 absorption band (53.65, 54.90 and 58.80 GHz). The last three were used to obtain temperature profiles from near the ground up to around 20 km altitude. Weighting functions appropriate to the channels are shown in Fig. 6.23 and the design of one channel in Fig. 6.24.

The main advantage possessed by a microwave radiometer compared with an infrared radiometer has already been mentioned, namely the much increased transparency of clouds at microwave frequencies. Another advantage occurs because radio techniques lead very readily to the achievement of very high spectral resolution, so that the weighting functions appropriate to the channels of a microwave instrument are essentially monochromatic ones (*cf.* §5.2).

The disadvantages of microwave compared with infrared measurements are (1) the field of view cannot be made so narrow because of the diffraction limit at the longer wavelength, (2) for the same accuracy of temperature measurement, the accuracy of radiance measurement has to be higher. A 1 K temperature difference leads to a change in radiance of $\sim \frac{1}{3}\%$ at 5 mm wavelength, $\sim 1\%$ at 15 μm and $\sim 4\%$ at 4.3 μm.

Fig. 6.25 gives some examples of comparisons between temperature retrievals from ITPR, NEMS, and temperature profiles from nearby radiosondes. All are for partially cloudy situations and the problems of retrieving accurate profiles under such circumstances are well illustrated. The retrieval errors from ITPR for case (d) are drastic enough for them to be easily recognized by internal consistency checks; this is probably not the case for the situation of Fig. 6.25(c). Fig. 6.26 shows the effect clouds associated with a tropical storm have on the NEMS and ITPR radiances. Even for such intense clouds the errors introduced into the NEMS radiances amount to less than 3 K. Considerations resulting from

Fig. 6.23 Weighting functions for the NEMS instrument on Nimbus 5: Zero surface reflectivity is assumed (after Staelin *et al.* 1972).

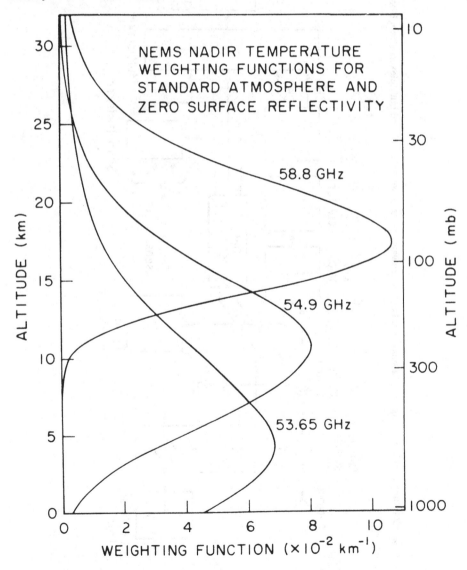

110

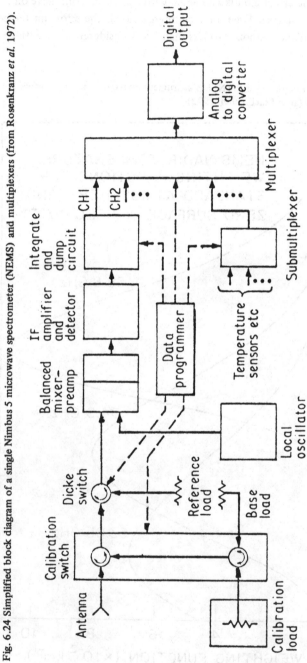

Fig. 6.24 Simplified block diagram of a single Nimbus 5 microwave spectrometer (NEMS) and multiplexers (from Rosenkranz *et al.* 1972).

the analysis of data of the kind shown in Figs. 6.25 and 6.26 have, therefore, led to a combination of infrared and microwave channels being mounted on the TIROS N operational satellite series beginning in 1978.

A scanning microwave spectrometer (SCAMS) was mounted on the Nimbus 6 satellite, its scan pattern being arranged to fit with that of HIRS. A combination of measurements from HIRS and SCAMS has been employed in studies of the accuracy of temperature sounding referred to in chapter 8. The microwave sounding unit (MSU) on the TIROS-N operational satellites is an adaptation of SCAMS. It has four channels within the 5 mm oxygen band at 50.3, 53.74, 54.96 and 57.05 GHz respectively. Its scan pattern fits with HIRS as shown in Fig. 6.10.

Fig. 6.25 Comparison between Nimbus 5 temperature soundings derived from NEMS (microwave) and ITPR (infrared) observations with conventional radiosonde observations during April 1972 (after Smith, Staelin and Houghton 1974).

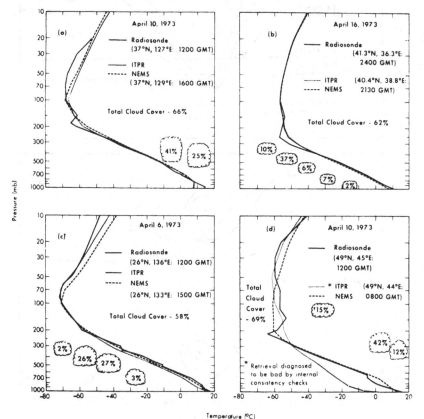

For the next generation of operational meteorological satellites for the late 1980s and 1990s, it is planned to build a more elaborate scanning microwave radiometer, the Advanced Microwave Sounding Unit (AMSU). This will have twenty channels, the fifteen lower frequency channels being provided by the US

Fig. 6.26 Brightness temperatures observed by NEMS (microwave) and ITPR (infrared) channels over a tropical storm in the South Pacific on 10 January 1973, 2200 GMT (top); mean difference between temperatures within the storm and those outside the storm's boundary (bottom) (after Smith, Staelin and Houghton 1974).

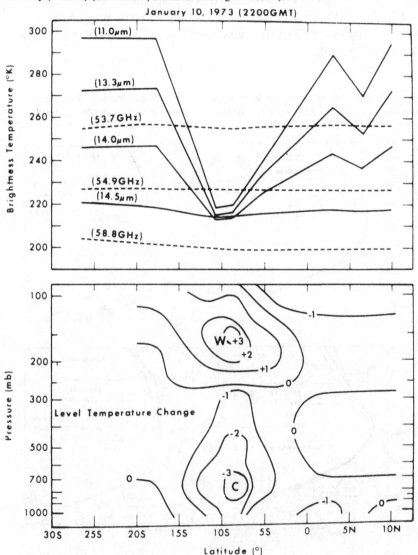

and the five high frequency channels by the UK. Table 6.6 lists the proposed channels which are also illustrated in Fig. 6.27; the weighting functions appropriate to the temperature sounding channels are shown in Fig. 6.28. Note that for some of the channels, several passbands symmetrically placed with respect to one or two line centres are included so as to increase the throughput of those channels. A temperature sensitivity of about 0.2 K is being set as a goal for the AMSU temperature sounding channels.

Table 6.6 *Advanced microwave sounding unit channel characteristics (from US National Oceanographic and Atmospheric Administration)*

Channel No. (Designation)	Passband Ctr. Frequency	Number of Passbands	Passband Bandwidth Per Band (MHz)	Beam Width (Degree)
1(W)	18.700 GHz	1	180	3.3
2(H)	23.800 GHz	1	380	3.3
3(W)	31.400 GHz	1	180	3.3
4(T)	50.300 GHz	1	180	3.3
5(T)	52.800 GHz	1	400	3.3
6(T)	53.330 GHz	1	400	3.3
7(T)	54.400 GHz	1	400	3.3
8(T)	54.940 GHz	1	400	3.3
9(T)	55.500 GHz	1	400	3.3
10(T)	57,290,344 MHz = f10	1	330	3.3
11(T)	f10 ±217 MHz	2	78	3.3
12(T)	f10 ±322.2 ±48 MHz	4	36	3.3
13(T)	f10 ±322.2 ±22 MHz	4	16	3.3
14(T)	f10 ±322.2 ±10 MHz	4	8	3.3
15(T)	f10 ±322.2 ±4.5 MHz	4	3	3.3
16(W)	89.0 GHz	1	6,000	1.0
17(H)	166.0 GHz	1	4,000	1.0
18(H)	183.31 ±1 GHz	2	1,000	1.0
19(H)	183.31 ±3 GHz	2	2,000	1.0
20(H)	183.31 ±7 GHz	2	4,000	1.0

Legend: Designation: (W) = window channel, (T) = temperature, (H) = humidity

6.10   Limb-sounding instruments

We have already mentioned in §5.1 that a different approach to the measurement of stratospheric temperatures from satellites involves the use of an instrument having a very narrow field of view which looks at the earth tangentially instead of straight down (Fig. 5.1). Height resolution is obtained by scanning the limb instead of spectroscopically, and so a fairly wide spectral bandwidth can be used, together with a fairly large telescope, to collect more energy and partially offset the effect of the very narrow field of view. The geometry of the limb path through the spherical atmosphere limits the vertical resolution of a limb radiance measurement to about 4 km (Fig. 5.5), and the horizontal resolution to about 500 km. The vertical resolution, therefore, is better than can be obtained with vertically viewing sounders while the horizontal resolution is on the whole not as good.

The first limb-sounding infrared radiometer (LRIR) to observe the atmosphere was flown on Nimbus 6 in 1975 (Gille *et al.* 1975). It possessed a viewing angle across the limb of about 1 mrad corresponding to a thickness of

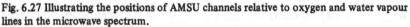

Fig. 6.27 Illustrating the positions of AMSU channels relative to oxygen and water vapour lines in the microwave spectrum.

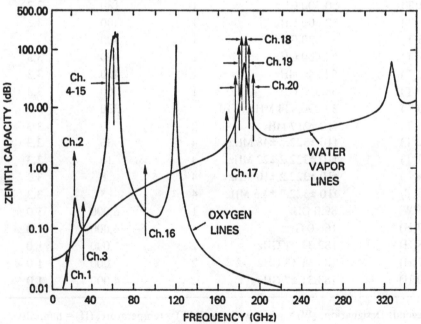

Fig. 6.28 Weighting functions for AMSU channels (from US National Oceanic and Atmospheric Administration).

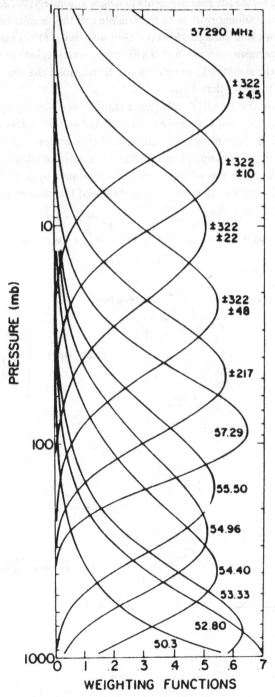

slice of atmosphere at the limb of about 2 km. It is not possible from geometrical considerations alone to locate the measurements in height absolutely, because the orientation of the instrument relative to the satellite and the satellite relative to the earth is not known well enough. However, Gille and House (1971) showed that by making radiance measurements in two different spectral regions of the 15 μm CO_2 band, a retrieval method can be devised which enables the absolute altitude to be inferred to better than 1 km.

   A similar instrument to LRIR called the Limb Infrared Monitor of the Stratosphere (LIMS) (Russell and Gille 1978) was mounted on the Nimbus 7 satellite launched in 1978. To achieve the required performance both instruments employed cadmium-mercury-telluride photoconductive detectors cooled to 63 K by a solid cryogen cooler consisting of solid methane surrounded by solid ammonia. The life of the cryogen in space limited the life of the instruments to

Fig. 6.29 Optical schematic of the LRIR and LIMS, from Gille *et al.* 1980.

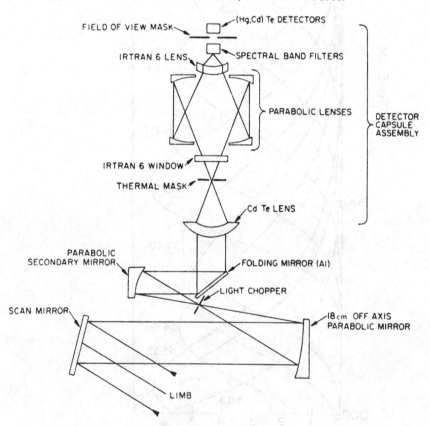

about 7 months. The optics of LIMS are shown in Fig. 6.29 and a list of the LIMS channels in Table 6.7. LRIR did not contain the NO_2 and HNO_3 channels. A temperature retrieval with the LIMS compared with a rocket sounding is shown in Fig. 6.30. Because of the long path being viewed through the atmosphere at the limb, significant temperature measurements up to altitudes of ~ 65 km are possible with these instruments.

A further infrared limb sounding instrument, the stratospheric and mesospheric sounder (SAMS) was also mounted on Nimbus 7. This was mainly directed at composition determination and is described in Ch. 10.

Fig. 6.30 Temperature obtained from LRIR measurements, using the direct first stage inversion technique (circles), compared to conventional rocketsonde and radiosonde measurements (solid line) from Wallops Island, Va. (37.85°N, 75.48°W) at 1800 UT on 19 November 1975. The LRIR observations were obtained at 37.95°N, 73.60°W at 1730 UT (from Gille *et al.* 1980).

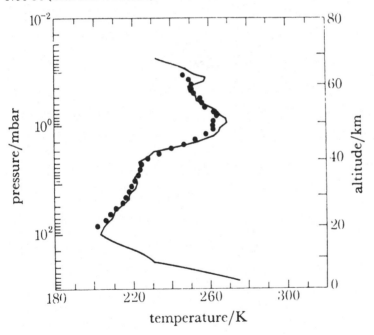

Table 6.7 *Characteristics of LIMS channels* (Gille *et al. 1980*).

Channel	Purpose	Band pass 50% relative response points (cm^{-1})	Field of view at limb (km)		Noise equivalent radiance $(watt/m^2\text{-sr})$
			Vertical	Horizontal	
1	NO_2	1580–1613	3.6	28	0.0006
2	H_2O	1396–1527	3.6	28	0.0021
3	O_3	947–1103	1.8	18	0.0035
4	HNO_3	859–900	1.8	18	0.0015
5	CO_2	595–739	1.8	18	0.005
6	CO_2	645–673	1.8	18	0.0013

7

RETRIEVAL THEORY

7.1   Introduction

The radiance emitted by an atmosphere is a relatively simple function
of the temperature distribution, as discussed in chapter 5, and a rather more
complicated function of the distribution of absorbers. We now address ourselves
to the question of the best method of interpretation of measurements of these
radiances in terms of understanding the physics and dynamics of the atmosphere.
In some cases it is possible to use the radiances directly as images, for there is
often much to be gained from simply looking at an atmosphere. This can be
done, of course, in spectral regions other than the visible; a good example is to
look at the earth's atmosphere in the region of the CO_2 15μ band during the
winter stratospheric disturbance known as 'sudden warming'. This can give
considerable insight into the nature of this particular phenomenon, as will be discussed further in Ch. 9. In many cases, however, it is necessary to derive distributions of actual values of such quantities as temperature or trace gas concentration in order to make proper use of the measurements. This is especially true
of composition, which is not related to the measurements in a simple way. The
distributions we require may be individual profiles, cross sections, three-
dimensional fields, or the time variation of these fields. The mathematics is
much the same in principle regardless of the dimensionality of the problems.

For the moment we will consider the problem of retrieving a temperature
profile from a set of radiances in a very simple set of circumstances, in order to
concentrate on the nature of the problem without extraneous difficulties. In the
case of a non-scattering atmosphere with a black lower boundary, the radiance
emitted vertically at the top of the atmosphere can be expressed as ($\S 5.2$):

$$L(v_i) = \int B(v_i, T(z)) K(z; \rho(z)) \, dz + B(v_i, T_0) \tau(0; \rho(z)) \qquad (7.1)$$

where $K(z; \rho(z))$ and $\tau(z; \rho(z))$ are the weighting function and transmittance respectively when the absorber density distribution is $\rho(z)$.

This is a solution to the *direct* problem of radiative transfer, i.e. 'given the distribution of temperature and absorber, calculate the radiation field'. The problem of remote sounding is an *inverse* problem, i.e. 'given the radiation field, find the distribution of temperature and absorber'. In fact, the problem is rather more difficult than this, because only the radiances in a restricted number of places and spectral regions are available. Usually this is called the *retrieval* problem, because in meteorological contexts 'temperature inversion' has a completely different meaning.

It might be thought that it is very easy to find a temperature profile which gives the same radiances as have been measured, and therefore the retrieval problem is trivial, and does not require much study. Unfortunately it is usually too easy to find solutions for any particular measurement, because the problem is ill-posed, and the solution is underconstrained. The real difficulty is not one of finding a solution, rather it is one of deciding which of an infinity of solutions is the most appropriate for the particular problem under consideration. In these circumstances, when the solution is not well-defined by the measurements, it is essential to make the best use of what information we have, and to treat approximations and short cuts with great care.

It is helpful to separate the problem conceptually into two parts, firstly the inverse problem of finding *a* solution, which may be easy if the relation is linear or nearly so, or difficult if it is nonlinear. Secondly, to find how far this solution can be altered and still be compatible with the measurements, i.e. to study uniqueness and noise sensitivity. The solution of the total problem includes criteria for the choice of the 'best' solution from all possible solutions. We will begin with a one-dimensional linear problem for which a mathematical solution is trivial, so that we can concentrate on uniqueness and noise. We will first describe some obvious but useless methods to highlight the difficulties, and then go on to explain the principles of estimation theory.

7.2   Simple solutions

7.2.1.   An exact solution to the linear problem

One of the simplest solutions to the radiative transfer equation is in the case of radiance $L(v)$ leaving the top of a non-scattering atmosphere in local thermodynamic equilibrium, when the optical depth is so great that the surface cannot be seen. In this case we may put

$$L(v) = \int_0^\infty B(v, z) \frac{d\tau(v, z)}{dz} \, dz \qquad (7.2)$$

where $B(v, z)$ is the Planck function at wavenumber v and height z, and $\tau(v, z)$ is the transmittance from height z to space. It is convenient to use

$z = -\ln(p)$, where p is pressure, as the height coordinate, so that the unit of z is the local scale height. This case is a close approximation to the real problem that arises in vertical temperature sounding, so it is convenient to use it to illustrate solution methods. If we assume that transmittance is independent of temperature, and that there are M measurements at a set of wavenumbers $\nu_1, \nu_2, ...\nu_M$, which are closely spaced, then

$$L_i = L(\nu_i) = \int_0^\infty B(\bar{\nu}, z) K_i(z)\, dz \quad i = 1, \ldots M \tag{7.3}$$

where $\bar{\nu}$ is a representative wavenumber and $K_i(z) = d\tau_i(\nu, z)/dz$. The equation is now linear in $B(\bar{\nu}, z)$, which can be regarded as the unknown profile. The radiance L_i to be measured is therefore a weighted mean of the Planck function profile with weighting function $K_i(z)$.

Solving equation (7.3) for $B(\bar{\nu}, z)$ is clearly an ill-posed or under-constrained problem because the unknown profile is a continuous function of height, and there is only a finite number of measurements. An obvious approach is to express $B(\bar{\nu}, z)$ as a linear function of M variables, b_j:

$$B(\bar{\nu}, z) = \sum_{j=1}^{M} b_j W_j(z) \tag{7.4}$$

where $W_j(z)$ is a set of representation functions such as polynomials or sines and cosines. Equation (7.3) becomes

$$L_i = \sum_{j=1}^{M} b_j \int_0^\infty W_j(z) K_i(z)\, dz = \sum_{1}^{M} A_{ij} b_j \tag{7.5}$$

thus defining the known square matrix **A**. For the M unknowns b_j there are M equations which in principle can be solved exactly. This approach was explored by Wark (1960) and Yamamoto (1961) in two of the earliest papers on the retrieval problem of remote sounding, who used linear interpolation between fixed levels and polynomials for $W_j(z)$. Unfortunately this simple method has a major drawback in that for practical situations it is usually ill-conditioned so that any experimental error in the measurements is greatly amplified, making the solution virtually useless. Furthermore, the more measurements that are made and thus the larger M becomes, the worse the problem becomes. This ill conditioning has been recognized for a long time in the mathematical and numerical literature as being a general feature of equations of this type, called Fredholm integral equations of the first kind.

The solution, obtained by solving equation (7.5) and substituting back in equation (7.6) may be written:

$$B(\bar{v}, z) = \sum_{ij} W_j(z) A_{ji}^{-1} L_i = \sum_i D_i(z) L_i \qquad (7.6)$$

where A_{ji}^{-1} is the jith component of the inverse matrix $\mathbf{A}^{-1}$. Equation (7.6) defines the 'contribution function' $D_i(z)$ as the contribution to the solution profile due to the measured radiance L_i. For an exact solution, i.e. one that gives the same theoretical radiances as the original profile, the contribution function obviously must satisfy

$$\int_0^\infty D_i(z) K_j(z)\, dz = \delta_{ij} \qquad (7.7)$$

where δ_{ij} is the Kroenecker delta. If there is an error ϵ_i in the measurement of L_i then there will be an error $\epsilon_i D_i(z)$ in the solution, thus ill conditioning is characterized by large values of $D_i(z)$. The noise variance $\sigma_B^2(z)$ of the solution at height z will be

$$\sigma_B^2(z) = \sum_i D_i^2(z)\, \sigma_i^2 \qquad (7.8)$$

where σ_i^2 is the variance of ϵ_i. If σ_i^2 is independent of i, then we may regard $\sum D_i^2(z)$ as a noise amplification factor. To illustrate this consider the synthetic set of weighting functions as shown in Fig. 7.1, from which the contribution functions $D_i(z)$ can be calculated for the exact solution, where the representation $W_j(z)$ is a set of polynomials in z. The sensitivity to noise is shown by the large values of $D_i(z)$ in Fig. 7.2. They are especially large outside the range of heights covered by the weighting functions, and cannot be shown on the same scale.

The representation so far is quite arbitrary. It is pertinent to ask if there is one for which the noise sensitivity is a minimum, because this might provide a logical choice for an exact solution. The answer is of course yes, and it may be found by jointly minimizing every element of $D_i^2(z)$ subject to the constraint expressed by equation (7.7). The best value of $D_i(z)$ is easily found using the calculus of variations to be

$$D_i(z) = \sum_j \left[\int_0^\infty K_j(z) K_i(z)\, dz \right]_{ij}^{-1} K_j(z) \qquad (7.9)$$

where the inverse is a matrix inverse. This solution may also be obtained by using the weighting functions, or any nonsingular linear combination of them, as a representation of the atmospheric profile. (Minimizing $D_i^2(z)$ in the derivation of equation (7.9) involved the implicit assumption that the experimental error in each of the components of the measurement has the same variance and is independent of the others. However, the result may be generalized to an arbitrary error covariance matrix). Mateer (1965) has studied this approach for the problem of Umkehr sounding for ozone, but it does not appear to have been used in the infrared remote sounding literature, although Smith (1970) has used weighting functions as a representation without commenting on the implications.

Fig. 7.3 shows how this choice has improved the contribution functions, compared with the polynomials used for Fig. 7.2. The extremely large value of $D_i(z)$ are no longer present, although $D_i(z)$ has increased in the middle range as a consequence of the constraint (equation 7.7) that the solution be exact. The absolute value of the sensitivity to noise is still unacceptably large, however, and an error of ϵ in a measurement will lead to errors of up to 15ϵ in the solution. This noise sensitivity is a feature of all exact solutions to

Fig. 7.1 A set of idealized weighting functions normalized to unit area.

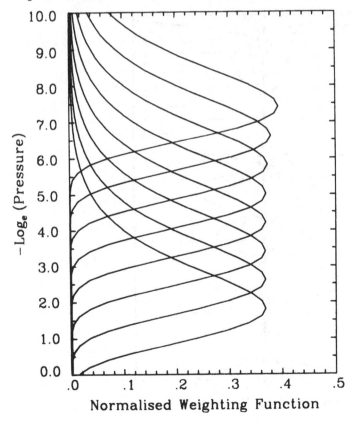

the inverse problem, whether linear or nonlinear, when there is significant overlap between the weighting functions. Qualitatively, it is difficult for the solution to follow noise in the observations without inducing oscillations, because if the solution has to be increased in the region of the peak of a weighting function in order to represent a positive error in the corresponding component of the measurements, there must be compensating decreases either side so that components due to the neighbouring weighting functions are un-affected. This means that the original increase must be larger than it otherwise would, and very much larger if the weighting functions are closely spaced. The compensating decreases will also induce compensating increases in the next component, and so on.

There is clearly no logical reason to try to find an exact solution. It is reasonable only to require that the solution lies within experimental error

Fig. 7.2 The contribution functions for an exact solution in terms of polynomials in z. The scale on the abscissa corresponds to the middle section of the graph. The scale for the outer sections is -1000 to $+1000$.

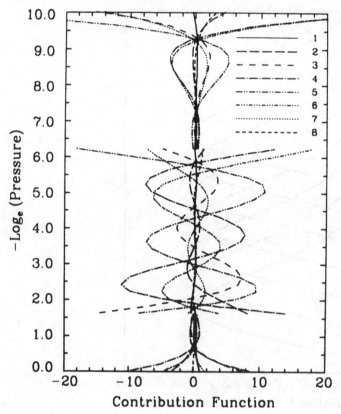

of the measurements. This gives us more freedom in choice of a solution, but it does raise the important question of what criteria should be used in this choice.

7.2.2. *Least squares solutions*

When faced with an ill-conditioned problem, a common approach is to use a least squares solution, that is, to solve overconstrained equations so that the sum of the squares of the differences between the measurements and the values corresponding to the solution is minimized. In the present case this may be done by representing the atmospheric profile with fewer functions than the number of components of the measurement

$$B(\bar{\nu}, z) = \sum_{j=1}^{N} b_j W_j(z), \ N < M \tag{7.10}$$

Fig. 7.3 The contribution functions for an exact solution in terms of the weighting functions.

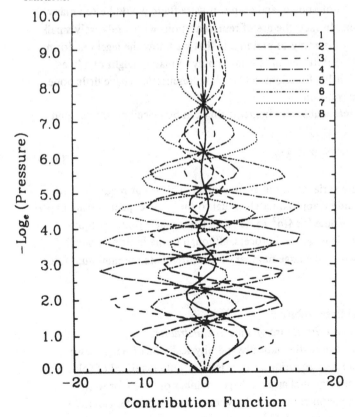

so that on substituting this into equation (7.3), the resulting set of equations for b_j is overconstrained:

$$L_i = \sum_{j=1}^{N} A_{ij} b_j, \; i = 1 \ldots N \qquad (7.11)$$

and must be solved by least squares. The solution $\hat{b}$ for which $\sum_i (L_i - \Sigma A_{ij} \hat{b}_j)^2$ is a minimum given by

$$\hat{b} = (A^T A)^{-1} A^T L. \qquad (7.12)$$

This solution is an improvement over an exact solution in the sense that its noise sensitivity is smaller, but there are still no reasonable criteria to determine such things as how many terms to use and what form of representation is suitable. The least squares method is really only appropriate if the algebraic form of the solution is known from sound physical reasoning, and the number of unknown parameters is considerably smaller than the number of independent measurements. It is not suitable for the atmospheric retrieval problem because the algebraic form is not known, and even if it were there would be too many unknown parameters. In fact, the use of representations with a relatively small number of parameters is a way of pretending that we know the algebraic form. The atmospheric temperature profile is not a polynomial in height or a linear combination of weighting functions, and such representations have little hope of providing a good solution.

The only reasonable requirements that can be placed on the squared deviation is that

$$\sum_i (L_i - \Sigma_j A_{ij} b_j)^2 \simeq M\sigma^2 \qquad (7.13)$$

where σ^2 is the noise variance in the measurement. This is not required to be a minimum, but should be approximately equal to the value expected from experimental error. It is possible for $M\sigma^2$ to be larger (or smaller) than the squared deviation found by a least squares process, although in the limit of a small number of unknowns and a large number of measurements the minimum value tends to $M\sigma^2$.

7.3   The general linear solution
7.3.1   Uniqueness and the nature of a priori constraints
It is clearly both mathematically and physically incorrect to solve an underconstrained set of equations. There is an infinite manifold of solutions which satisfy them exactly, and an even larger number of solutions which satisfy them within experimental error. It is necessary to consider carefully

whether the problem to be solved is realistic, and in what sense a solution exists
The problem of retrieval may be restated as:

> 'Given the measured radiances **L**, the statistics of the experimental
> error ϵ, the instrumental weighting functions **K**, and any other relevant
> information, what can be said in a physically meaningful way about
> the unknown **B**?'

Some possible approaches to this question are:

(1)   Characterize the class of all possible solutions for example by means
       of its probability density function, and choose some property of
       the distribution such as its mean or mode as the solution.

(2)   Choose some specified function of the unknown profile, for example
       the thickness between pressure surfaces, and characterize its probab-
       ility density function as in (1).

(3)   Find a smoothed version of the true profile, with a known smoothing
       function, by linear combination of the measurements.

To carry out either (1) or (2) 'other relevant information' in the form of constraints on the solution profile beyond those provided by the measurement is
needed, in order that the error variance of the solution be finite. If no contraints
are available, or the constraints are insufficient, then only approach (3) is valid.
A priori constraints may arise from many sources, some of which are described
below. Probably the most enlightening approach to understanding is to regard
them as 'virtual measurements', that is, as being of the same nature as real
measurements. They say something about the unknown profile just as the
measurements do, and together with the measurements they determine
whether the problem is well-posed. This is the same philosophy as that employed
by Franklin (1970) in a context of more general applicability.

A linear constraint is one which takes the same mathematical form as
a linear direct measurement, i.e., it gives a value for a known linear function
of the unknown profile, together with an error covariance matrix for this value.
If both the constraints and the direct measurements are linear, then so is the
problem. If either or both are nonlinear, it is a nonlinear problem.

Great care must be exercised in selecting the constraints to be applied,
so that they represent the best possible *a priori* information available. It is
possible for a constraint to be too tight, too loose, or simply wrong. A good
example of the latter is using inappropriate statistics or an unsuitable representation, and an example of the other two possibilities is the use of a wrong
size of smoothing parameter for the Twomey-Tikhonov solution (§ 7.5.2).

Many of the constraints that have been used in the literature fall into
the class of linear constraints. Some of them are described below to help the
reader recognize linearity:

Statistics. The climatological mean profile and its covariance may be regarded as a measurement and its uncertainty.

Forecast profile. The forecast profile and its error covariance may similarly be regarded as a measurement and its uncertainty. However, the error covariance is often not known very well, so must be over-estimated.

Twomey–Tikhonov. The virtual measurement is zero or an *a priori* guessed profile. The error covariance is proportional to Twomey's constraint matrix H, often taken to be a unit matrix.

Linear representations. It is less obvious that a linear representation with a finite number of terms is a linear constraint as defined above. The representation may be extended to be a complete one with an infinite number of terms by including any infinite set of orthogonal functions which are all orthogonal to the original representation. The coefficients of the extended representation are then known linear functions of the unknown profile. The linear constraint states that the values and variances of the extra terms are all zero, whilst the values of the coefficients of the original representation are zero with infinite variances.

Discretization. It is usual to express continuous functions such as the unknown profile and the weighting functions in a discrete form with a relatively large number of levels (typically 50–200) in order to simplify the numerical calculations, and so that the algebra of vectors and matrices may be used instead of Hilbert space. This is of course a linear representation and may be regarded as a linear constraint exactly as above. The discretization must have high enough spatial resolution so that the implied assumptions are valid.

The general approach to the linear problem should now be clear. There exists a set of direct measurements which alone lead to an ill-posed problem. They must be supplemented with enough virtual measurements to make the problem well posed. This may be done in two stages, discretization and *a priori* constraints, although the distinction is not essential. Each measurement, direct or virtual, is in the form of an estimate of a linear function of the profile and a covariance. They are combined in the usual way of combining measurements, i.e., by weighting independent estimates with inverse covariances.

In the following sections a discrete version of the problem is used for the reasons described above. Equation (7.3) will be replaced by

$$\mathbf{y} = \mathbf{K}\mathbf{x} \tag{7.14}$$

where $\mathbf{y}$ is a vector of qualities to be measured and $\mathbf{x}$ is the unknown profile. The matrix $\mathbf{K}$ is a discrete version of the weighting functions, and the integral becomes a matrix multiplication. The change from L to y and from B to x has been made for consistency with literature in other fields, and is in accord with usual mathematical practice. Normally it will not be necessary to distinguish between y and a measurement of y whose value is y + ϵ, where ϵ is experimental error. It should be clear from the context which is intended.

7.3.2 Combination of observations

The reader should be familiar with the usual way of combining two independent measurements x_1 and x_2 of a scalar quantity x by taking a weighted average, with the reciprocal of the square of the standard deviations as weights:

$$\hat{x} = (1/\sigma_1^2 + 1/\sigma_2^2)^{-1} (x_1/\sigma_1^2 + x_2/\sigma_2^2) \tag{7.15}$$

where $\hat{x}$ denotes the combined estimate of x. The variance of x is

$$\hat{\sigma}^2 = (1/\sigma_1^2 + 1/\sigma_2^2)^{-1}. \tag{7.16}$$

This generalizes to vectors in a straightforward manner. If there are two measurements of the vector x, namely x_1 and x_2, with error covariances S_1 and S_2, then the estimate x is

$$\hat{x} = (S_1^{-1} + S_2^{-1})^{-1} (S_1^{-1} x_1 + S_2^{-1} x_2) \tag{7.17}$$

with covariance

$$\hat{S} = (S_1^{-1} + S_2^{-1})^{-1}. \tag{7.18}$$

The notation S is used rather than Σ for covariance matrices to avoid confusion with summation. In the case of the retrieval problem we often have a virtual measurement x_0 with covariance S_x and a direct measurement of y = Kx with error covariance S_ϵ. In this case the estimate is

$$\hat{x} = (S_x^{-1} + K^T S_\epsilon^{-1} K)^{-1} (S_x^{-1} x_0 + K^T S_\epsilon^{-1} y) \tag{7.19}$$

with covariance

$$\hat{S} = (S_x^{-1} + K^T S_\epsilon^{-1} K)^{-1} \tag{7.20}$$

These two equations are easily related to equations (7.17) and (7.18) by noting that D y is an estimate of x with inverse covariance $S_D^{-1} = K^T S_\epsilon^{-1} K$, where D is any exact solution (e.g. equation 7.9) such that $\mathbf{K}\,\mathbf{D} = \mathbf{I}$, the unit matrix. In these terms equation (7.19) becomes

$$\hat{x} = (S_x^{-1} + S_D^{-1})^{-1}\,(S_x^{-1}\,x_0 + S_D^{-1}\,\mathbf{D}\,y) \tag{7.21}$$

which is of the same form as equation (7.17). Note that S_D^{-1} is singular, so that S_D itself cannot be evaluated. The variances of the elements of $\mathbf{D}y$ are infinite.

Equations (7.19) and (7.20) for the best estimate of the profile and its covariance are useful for understanding the solution, but they are not useful for practical computation because they require the inversion of relatively large matrices, or equivalently, the solution of large systems of linear equations. By making use of the obvious matrix identity

$$\mathbf{K}^T(\mathbf{I} + S_\epsilon^{-1}\,\mathbf{K}\,S_x\,\mathbf{K}^T) = \mathbf{K}^T + \mathbf{K}^T\,S_\epsilon^{-1}\,\mathbf{K}\,S_x\,\mathbf{K}^T = (\mathbf{I} + \mathbf{K}^T\,S_\epsilon^{-1}\,\mathbf{K}\,S_x)\,\mathbf{K}^T, \tag{7.22}$$

equation (7.19) takes the form

$$\hat{x} = x_0 + S_x\,\mathbf{K}^T\,(\mathbf{K}\,S_x\,\mathbf{K}^T + S_\epsilon)^{-1}\,(y - \mathbf{K}\,x_0), \tag{7.23}$$

where there is only one relatively small matrix to be inverted. Also using equation (7.22), the expression (7.20) for the covariance of x can be reduced to a computationally simpler form, although it is algebraically more complex:

$$\hat{S} = S_x - S_x\,\mathbf{K}^T\,(\mathbf{K}\,S_x\,\mathbf{K}^T + S_\epsilon)^{-1}\,\mathbf{K}\,S_x \tag{7.24}$$

Fig. 7.4 gives an example of the contribution functions calculated according to equation (7.23), where S_ϵ has been taken to be $0.1\,\mathbf{I}$ and S_x is a covariance matrix computed from a sample of mid and high latitude rocket soundings. Diagonal elements of S_x are of the order of magnitude of 100 [mWm^{-2} ster^{-1} (cm^{-1})$^{-1}$]^2 It can be seen that the noise amplification problem is considerably reduced in comparison with Figs. 7.2 and 7.3.

Equations equivalent to (7.23) have been derived or applied by Westwater and Strand (1968), Turchin and Nozik (1969), Rodgers (1970, 1971), and DeLuisi and Mateer (1971), amongst others.

A derivation of these equations in a more general form should give some insight into the nature of the solution. The two basic ways of combining noisy measurements are to find the most likely value or to find the expected value. In general they will lead to different solutions (the 'mode' or the 'mean'), but if the error statistics are Gaussian the maximum likelihood and expected value combinations are identical.

The multivariance Gaussian distribution may be written as

$$P(\mathbf{x}) = \frac{1}{(2\pi)^{n/2}\det(\mathbf{S})^{1/2}} \exp\left\{-\tfrac{1}{2}(\mathbf{x}-\mathbf{x_0})^T\mathbf{S}^{-1}(\mathbf{x}-\mathbf{x_0})\right\} \qquad (7.25)$$

where $P(\mathbf{x})$ is the probability density function (pdf) of the random vector $\mathbf{x}$, whose dimension is n, $\mathbf{x_0}$ is the expected value of $\mathbf{x}$, $E\{\mathbf{x}\}$, and $\mathbf{S}$ is the $n \times n$ covariance matrix, i.e.,

$$\mathbf{S} = E\{(\mathbf{x}-\mathbf{x_0})(\mathbf{x}-\mathbf{x_0})^T\}. \qquad (7.26)$$

Let there be l independent measurements y_i, $i = 1 \ldots l$, of linear functions of the unknown profile $\mathbf{x}$:

$$y_i = \mathbf{K}_i\mathbf{x}, \qquad (7.27)$$

Fig. 7.4 The contribution functions for the linear solution with a statistical constraint.

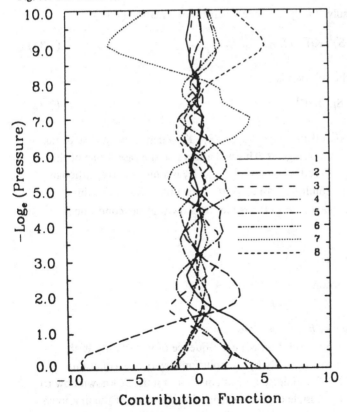

each with error covariance S_i. (So far we have considered the case where $l = 2$.) The most likely value of x given the set data y_i is that which maximizes the conditional probability density function $P(x|y_1, y_2 \ldots y_l)$. By use of Bayes theorem this may be written as

$$P(y_1, y_2 \ldots y_l | x) P(x)/P(y_1, y_2 \ldots y_l). \qquad (7.28)$$

If the measurements are independent we may further expand this as

$$
\begin{aligned}
& P(y_1|x)P(y_2|x) \ldots P(y_l|x)P(x)/P(y_1)P(y_2) \ldots P(y_l) \\
& = P(x) \prod_i P(y_i|x)/P(y_i)
\end{aligned}
\qquad (7.29)
$$

If nothing is known about x other than the values of y_i, then $P(x)$ is a constant. To find $\hat{x}$, the most likely value of x, we maximize with respect to x, so the terms $P(y_i)$ are also constants. Thus we maximize $\prod_i P(y_i|x)$, or minimize minus its logarithm:

$$\frac{\partial}{\partial \hat{x}} \sum_i (y_i - K_i\hat{x})^T S_i^{-1} (y_i - K_i\hat{x}) = 0 \qquad (7.30)$$

This may easily be solved to give a generalization of equation (7.19):

$$\hat{x} = \left(\sum_i K_i^T S_i^{-1} K_i\right)^{-1} \left(\sum_i K_i^T S_i^{-1} y_i\right). \qquad (7.31)$$

The covariance of the solution is

$$\hat{S} = \left(\sum_i K_i^T S_i^{-1} K_i\right)^{-1}. \qquad (7.32)$$

The solution only exists if the matrix $\sum_i K_i^T S_i^{-1} K_i$ which is the inverse of the covariance matrix $\hat{S}$, is nonsingular. This will always be the case if one of the measurements is of x directly, i.e., if one of the K_i is a unit matrix, although this is not a necessary condition. An *a priori* constraint on x ensures this.

It is left as an exercise to the reader to show that the conditional expectation

$$\hat{x} = E\{x|y_1, y_2 \ldots y_l\} \qquad (7.33)$$

leads to the same solution.

7.3.3. *The accuracy of linear solutions*

Uncertainty in the solution is a consequence of uncertainty in the measurements and is given formally by the solution covariance, equations (7.20), (7.24) or (7.32). To understand this equation it must be known how to interpret it in terms of possible deviations from the true value of the unknown profile. The diagonal of the covariance matrix contains the variances, or squares of standard deviations, of the individual components of the profile **x**, thus giving the simplest measure of accuracy. Westwater and Strand (1968) and Rodgers (1970) have used the residual variance, i.e. $\hat{S}_{jj}$, as a measure of the accuracy of the solution, and have shown that the linear solution given by equation (7.23) is the one for which residual variance is a minimum. The off-diagonal elements of $\hat{S}$ are non-zero, so that there are correlations between the errors at different levels, and merely stating that $x_j \pm \hat{S}_{jj}^{1/2}$ is known is an understatement of the knowledge of **x**. Another approach is to find some way of expressing the error estimate as a sum of individual components, such that the components are uncorrelated with each other. That is, it is desirable to express the error in $\hat{x}$ in the form

$$\hat{x} - x = \sum_i \epsilon_i v_i \qquad (7.34)$$

where the ϵ_i are *independent* random variables, and the vectors v_i are some known set of functions. This can be done by diagonalizing $\hat{S}$, i.e. by finding its eigenvectors and values:

$$\hat{S} V = V \Lambda \text{ or } \hat{S} v_i = \lambda_i v_i \qquad (7.35)$$

where the eigenvectors **v**, of **S** form the columns of the matrix **V**, and the eigenvalues λ_i form the diagonal elements of the diagonal matrix Λ. The eigenvectors may be normalized so that

$$V^T V = V V^T = I. \qquad (7.36)$$

Therefore

$$V^T \hat{S} V = \Lambda \text{ and } \hat{S} = V \Lambda V^T \qquad (7.37)$$

If $\hat{x}$ has error covariance $\hat{S}$, then $V^T \hat{x}$ has error covariance $V^T \hat{S} V$, i.e., Λ. Thus the errors in the quantities $v_i^T \hat{x}$ are independent, because Λ is diagonal. Now $a_i = v_i^T \hat{x}$ is the coefficient of v_i in an expansion of $\hat{x}$ in terms of the eigenvectors, i.e.,

$$\hat{x} = \sum_i a_i v_i \quad \text{where} \quad a_i = v_i^T x \qquad (7.38)$$

and the error variance in a_i is λ_i. Therefore we may say that the error in $\hat{x}$ is of the form

$$\hat{x} - x = \sum_i \epsilon_i v_i \qquad (7.39)$$

where the ϵ_i's are independent random variables with variance λ_i.

As an illustration the first eight eigenvectors and values for the case
of the synthetic weighting functions of Fig. 7.1 have been computed with experimental error covariance $S_\epsilon = 0.1\ I$ and the same covariance matrix S_x as was
used for Fig. 7.4. The eigenvalues are given in Table 7.1, and the eigenvectors
in Fig. 7.5.

There is no very clear cut relationship between the size of the eigen-
value and the shape of the eigenvector, except that the larger eigenvalues tend to
be associated with uncertainties outside the range of the weighting functions.
As a consequence the detailed shapes of the eigenvectors will depend on the
range of heights under consideration.

7.4   Information content of a measurement

The name 'information content' has been used to cover a variety of
concepts in the remote sounding literature. Mateer (1965) used it to describe
the number of statistically independent quantities that a sounding system can
measure, in the sense discussed above in § 4.3. Westwater and Strand (1968)
used 'information content' to describe the reduction in total variance due to
a measurement, where total variance is the sum of the diagonal elements of a
covariance matrix. Twomey (1970), used the name in a paper discussing the
number of independently measurable quantities, in which he pointed out that
increasing the number of spectral intervals does not necessarily increase the
information content significantly unless there is a corresponding reduction of
noise, because of the inherent resolution of the weighting functions. Peckham
(1974) used the definition employed in information theory in a paper optimiz-
ing the configuration of a remote sounding radiometer.

All of the above are useful concepts, but the name 'information content' should be reserved for the meaning assigned to it by information theory.
Qualitatively it is reasonable to ask that the information content of a measure-

Table 7.1 *Eigenvalues of the matrix* $\hat{S}$.

n	λ_n	$\lambda_n^{1/2}$
1	469.1	21.7
2	274.1	16.6
3	83.9	9.2
4	47.0	6.9
5	39.5	6.3
6	35.5	6.0
7	21.2	4.6
8	18.8	4.3

ment should be a measure of the factor by which the uncertainty in the value of a quantity is decreased by the act of measuring it. In fact the definition used in information theory (Shannon and Weaver 1949) may be regarded roughly as the number of bits (binary digits) needed to represent the number of distinct measurements that could have been made.

Information is the measure of change in a quantity called *entropy*. The entropy of a probability density function P(x) is defined as

$$H(P) = - \int P(x) \log_2 P(x) \, dx \, . \tag{7.40}$$

The information content of a measurement is defined as

$$H(P_1) - H(P_2) \tag{7.41}$$

Fig. 7.5 Eigenvectors of the solution error covariance for the linear solution with a statistical constraint.

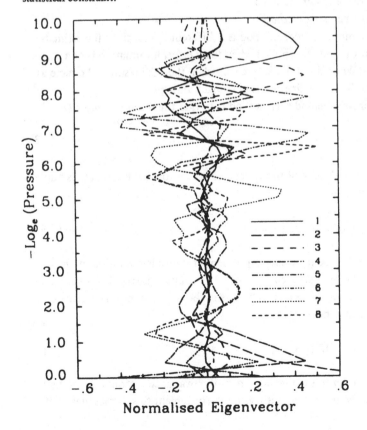

where P_1 is the *a priori* probability density function of the unknown **x**, and P_2 is its pdf after the measurement has been made. It is easy to show that if $P(\mathbf{x})$ is a Gaussian distribution with covariance S, then

$$H(P) = \tfrac{1}{2} \log_2 |S| = \log_2 |S|^{1/2} \qquad (7.42)$$

Contours of probability density in the case of a Gaussian pdf may be described as *n*-dimensional ellipsoids whose axes are in the directions of the eigenvectors of S, with lengths equal to the square roots of the corresponding eigenvalues. The determinant of a matrix is equal to the product of the eigenvalues, therefore $|S|^{1/2}$ is proportional to the volume inside such an ellipsoidal contour. So the information content of a measurement is, on Shannon's definition, equal to the logarithm of the ratio of the 'volumes of uncertainty' before and after the measurement.

7.5   A handful of linear methods
7.5.1   *Minimum variance method*
The minimum variance method is a different approach to the estimation problem that can only be applied when the *a priori* information is of a statistical nature. When the statistics are Gaussian, then the result is the same as the maximum likelihood estimator, as one might anticipate.

The principle behind this method is to find the linear predictor **D**,

$$\hat{\mathbf{x}} = \mathbf{D}\,\mathbf{y} \qquad (7.43)$$

such that the expected value of the variance of the error in the estimate is minimum, i.e. we minimize

$$E\{(\mathbf{x} - \hat{\mathbf{x}})^T(\mathbf{x} - \hat{\mathbf{x}})\}. \qquad (7.44)$$

This is of course the classical problem of multiple regression, with the elements of **D** as the regression coefficients. The solution is straightforward, and is left as an exercise for the reader. It results in the so called 'normal equations', which in this case may be written

$$\mathbf{D} = E\{\mathbf{x}\,\mathbf{y}^T\}\;[E\{\mathbf{y}\,\mathbf{y}^T\}]^{-1} \qquad (7.45)$$

The expected values in this expression are both covariance matrices. They may be evaluated exactly as written, but it is more convenient to subtract means of x and y to obtain the covariance matrices that have been used in previous sections. If $y = K x + \epsilon$ then it is easily shown that

$$E\{x\,y^T\} = S_x\,K^T \quad \text{and} \quad E\{y\,y^T\} = S_y = K\,S_x\,K^T + S_\epsilon \qquad (7.46)$$

Thus the minimum variance solution leads to

$$\hat{x} = S_x\,K^T(K\,S_x\,K^T + S_\epsilon)^{-1}\,y \qquad (7.47)$$

which is identical to equation (7.23) for the maximum likelihood solution.

The minimum variance solution may be used in this form, or used directly in its multiple regression form (eqn. 7.45). If there is a sufficiently large sample of cases where y has been measured by remote soundings, along with independent direct measurements from radiosondes or rockets, then **D** can be estimated entirely from experimental data without the need to know the weighting functions. This method has been applied in practice to SIRS on NIMBUS 3 (Smith, Woolf and Jacob, 1970), and is in use for routine soundings from the TIROS Operational Vertical Sounder (TOVS).

7.5.2   The Twomey-Tikhonov method

The first method applied to the retrieval problem in which due consideration was given to the estimation problem was that of Twomey (1963) and Tikhonov (1963). The principle enunciated by Twomey in his first presentation of this method is to find an estimate $\hat{x}$ such that the calculated measurement $\hat{y} = K\,\hat{x}$ differs from the actual measurement by exactly the experimental error

$$(y - K\,\hat{x})^T(y - K\,\hat{x}) = M\sigma^2 \qquad (7.48)$$

and such that $\hat{x}$ minimizes a given quadratic form:

$$(\hat{x} - x_0)^T H(\hat{x} - x_0), \qquad (7.49)$$

where **H** is chosen according to the problem in hand. Twomey suggests various possibilities including minimizing the variance of x about x_0 (**H** = **I**), minimizing the curvature, or second differences of **x**, etc.

When the minimization problem is formulated using a Lagrangian multiplier:

$$\frac{\partial}{\partial \hat{x}}\{(\hat{x} - x_0)^T H(\hat{x} - x_0) + \gamma(y - K\,\hat{x})^T(y - K\,\hat{x})\} = 0 \qquad (7.50)$$

it is exactly the same form as the maximum likelihood solution if we interpret x_0 and H^{-1} as the *a priori* constraint and its covariance, and γI as the inverse error covariance S_ϵ^{-1}.

Twomey finds that it is algebraically very difficult to determine the Lagrangian multiplier γ from his constraint, so he proposes that it be determined empirically. The maximum likelihood method gives an interpretation which allows one to choose a reasonable value for γ.

This method was arrived at independently by Tikhonov, who called it 'regularization'. The case where H^{-1} is a statistical covariance matrix is often called 'statistical regularization' in the Russian literature.

7.5.3   Truncated orthogonal expansion methods

Occasionally methods have been proposed that involve finding an exact or least squares solution using some kind of orthogonal expansion. The commonest sets of functions are the eigenvectors of $K^T K$ and the empirical orthogonal functions, or eigenvectors of S_x. For example, see Sellers and Yarger (1969) and Mateer (1965).

Such methods are nonoptimum for the reasons discussed in § 7.2.2 although a method using empirical orthogonal functions may be nearly optimum, but it seems unnecessary to use this approach when the algebra and computation for an optimum method is so little different. However, a situation where a truncated empirical orthogonal function representation may be used to advantage is in improving computational and storage efficiency on small computers. An expansion may be used instead of a high-resolution discretization to represent the profile, provided that the number of terms used is somewhat greater than the number of spectral intervals. For example, if the instrument has 10 spectral intervals, it may be found that a representation in terms of perhaps 20 empirical orthogonal functions is as good as a discretization at 100 levels.

7.5.4   Sequential estimation

The optimal estimation equations

$$\hat{x} = x_0 + S_x\, K^T\, (K\, S_x\, K^T + S_\epsilon)^{-1}\, (y - K\, x_0) \tag{7.51}$$

$$\hat{S} = S_x - S_x\, K^T\, (K\, S_x\, K^T + S_\epsilon)^{-1}\, K\, S_x \tag{7.52}$$

are of general applicability and there are many ways of using them.

For example it is not necessary to perform a matrix inversion (or solve simultaneous linear equations explicitly) because these equations can be

used sequentially. We may treat the measurement y as a set of scalars
$y_i, i = 1 \ldots M$, and perform the following operations:

$$S_0 = S_x$$
for $i = 1$ to M:
$$\left\{ \begin{array}{l} x_i = x_{i-1} + S_{i-1} k_i (y_i - k_i^T x_{i-1})/(k_i^T S_{i-1} k_i + \sigma^2) \\ S_i = S_{i-1} - S_{i-1} k_i k_i^T S_{i-1}/(k_i^T S_{i-1} k_i + \sigma^2) \end{array} \right\} \qquad (7.53)$$
$$\hat{x} = x_M; \hat{S} = S_M$$

The column vector k_i is the ith weighting function, or the ith row of K. Each
element of y is treated separately and at each stage a new estimate of x is made,
using it as the *a priori* value for the next stage. The matrix inverse becomes a
scalar reciprocal, and the final value of $\hat{x}$ is identical to that obtained from
original formula. In the linear case this has few advantages over the original
method - the number of arithmetic operations is similar - although the computer
program is shorter, and it is useful for very small machines. However, in the
nonlinear problem (§ 7.7) it can speed up convergence if the measurements
are used in the right order – the more linear ones first.

  The concept of sequential estimation allows one to use the continuity
that exists along a subsatellite track to improve the accuracy of a retrieval. It
is possible in principle to retrieve a whole orbit's or even whole day's data in
one operation, using equations (7.19) and (7.20) where x now refers to the two
or three-dimensional distribution of temperature, but in practice it is difficult
to estimate S_x, and the size of the matrices involved is prohibitive. However,
if something is known about how the profile changes in the horizontal a better
a priori estimate can be made using the retrieval previous in time and therefore
close in space.

  The simplest situation is to assume that the profiles follow a simple
first-order stochastic process:

$$x_n = \alpha x_{n-1} + \beta \qquad (7.54)$$

where x_n is the profile at position n, α is a known matrix and β is a random
vector with zero mean and known covariance S_β. A slightly better assumption
might be

$$(x_n - \bar{x}_n) = \alpha(x_{n-1} - \bar{x}_{n-1}) + \beta \qquad (7.55)$$

where $\bar{x}_n$ is a climatology or forecast, i.e. the quantity that would be used for
x_0 if separate rather than sequential retrievals were being performed.

If the retrieval at position $n - 1$ has given an estimate $\mathbf{x}_{n-1}$ with co-variance $\hat{\mathbf{S}}_{n-1}$, then the stochastic process may be used to obtain an *a priori* estimate for $\mathbf{x}_n$:

$$\mathbf{x}_n = \bar{\mathbf{x}}_n + \alpha(\hat{\mathbf{x}}_{n-1} - \bar{\mathbf{x}}_{n-1}) + \beta \tag{7.56}$$

with covariance

$$\mathbf{S}_n = \alpha\,\hat{\mathbf{S}}_{n-1}\,\alpha^T + \mathbf{S}_\beta \tag{7.57}$$

These estimates are then combined with the observation at position n to obtain the estimate $\hat{\mathbf{x}}_n$, and its covariance $\hat{\mathbf{S}}$. This process gives a one-sided smoothing, but it can be improved by repeating the process along the orbit in the reverse direction, and taking a proper weighted mean of the forward estimate $\hat{\mathbf{x}}_n$ and the backward *a priori* estimate $\mathbf{x}_n$. (Averaging the forward estimate and the backward estimate uses the measurement twice.)

Estimation of the matrices α and $\mathbf{S}_\beta$ is a difficult problem. In principle they should be obtainable from appropriately spaced *in situ* measurements, but in practice these are hard to obtain. It may be possible to construct $\mathbf{S}_\beta$ from statistics of the zonal wind in the case of sounding temperature from a polar orbiting satellite. A crude method which is nevertheless useful is to assume $\alpha = \gamma\mathbf{I}$, where γ is an empirically chosen constant close to unity, but a little smaller, in which case it is necessary to have

$$\mathbf{S}_\beta = (1-\gamma^2)\,\mathbf{S}_x \tag{7.58}$$

for consistency with the global statistics.

7.6   Solvable problems - the Backus-Gilbert approach

If there are not enough measurements and constraints to make the problem well-posed, or if those that there are do not reduce the solution error covariance $\hat{\mathbf{S}}$ sufficiently, then the original problem is not solvable, and we must look for other problems that are. This is the basic philosophy adopted by Backus and Gilbert (1970) in their approach to the problem of remotely sounding the structure of the solid earth with seismic waves, which was applied to the atmospheric remote-sounding problem by Conrath (1972).

In the case of the linear problem, all we can reasonably do with the measurements is to take linear combinations of them. Every linear combination of measurements corresponds to a linear function of the unknown profile.

The essence of the Backus-Gilbert method is to control the shape of this
linear function so that it corresponds to some meaningful quantity.

A typical useful shape for the case of atmospheric sounding might
be something like a 'boxcar' function so that the corresponding combination of measurements approximates mean temperature over a layer,
which is proportional to the thickness of the layer. However most of the
development of this method has been directed towards approximating a
delta function, resulting in a singly peaked function with a width that
may be specified by the user. The result is that the unknown profile may
be represented at a range of spatial resolutions, but not at infinite resolution.
The highest resolution attainable depends on the set of weighting functions. The
error in a profile seen at finite resolution is of a different kind from one seen at
infinite resolution but with measurement error. For example, the difference
between the finite resolution profile and the original profile will depend on the
curvature of the original profile, whilst the errors in the infinite resolution
profile depend only on the error covariances and not on the original profile.

Of course experimental noise is important for the Backus-Gilbert
method, and as one might expect, it is related to resolution. A high-
resolution profile will have high noise, and a low-resolution profile will
have low noise. One can trade off resolution against noise, to obtain
the best compromise for any particular application.

In order to approximate a delta function by a linear combination
of weighting functions some parameter is needed which measures how
good the approximation is and which may be minimized for a best fit. Such
a quantity is the *spread*, which is defined as follows: the spread $S(x)$ of a
function $A(z)$ of height z about a height x is

$$S(x) = 12 \int (x - z)^2 A^2(z) dz .$$ (7.59)

The normalizing factor 12 is included so that a boxcar function centered
at x will have a spread equal to its width. The function $A(z)$ must have
unit area:

$$\int A(z) dz = 1$$ (7.60)

Backus and Gilbert tried several possible definitions of spread before
choosing the one given above. Other definitions, such as the 'radius of
gyration' of $A(z)$ produced significant negative excursions in the solution,
or had other similar disadvantages. In this section we will revert to our
original notation in which quantities are treated as functions of height
rather than in a discretized form as a vector.

A solution profile $B_s(x)$ seen at finite resolution is sought which is given by a linear function of the measurements:

$$B_s(x) = \sum_i D_i(x) L_i \qquad (7.61)$$

Substituting for L_i from equation (7.3) gives

$$B_s(x) = \int \sum_i D_i(x) K_i(z) B(z) dz , \qquad (7.62)$$

showing that $B_s(n)$ is a smoothed version of $B(z)$ with a smoothing function:

$$A_x(z) = \sum_i D_i(x) K_i(z) \qquad (7.63)$$

We wish to choose $D_i(x)$ so that $A_x(z)$ has the smallest possible spread. The spread of this function is found using equation (7.59) to obtain

$$S(x) = \sum_{jl} D_j(x) D_l(x) Q_{jl}(x) \qquad (7.64)$$

where

$$Q_{jl}(x) = 12 \int (x - z)^2 K_j(z) K_l(z) dz , \qquad (7.65)$$

and

$$\sum_j D_j(x) = 1 \qquad (7.66)$$

because $A(z)$ and $K(z)$ are both normalized to unit area. We therefore minimize $S(x)$ with respect to $D_k(x)$ subject to the unit area constraint:

$$\frac{\partial}{\partial D_k(x)} \left\{ \sum_{jl} D_j(x) D_l(x) Q_{jl}(x) + \lambda \sum_j D_j(x) u_j \right\} = 0 \qquad (7.67)$$

where λ is an undetermined multiplier and $u_j = 1$ for mathematical convenience. This gives:

$$\sum_j D_j(x) Q_{jk}(x) + \lambda u_k = 0 \qquad (7.68)$$

Therefore, in matrix notation

$$\mathbf{D}^T(x) = - \lambda \mathbf{u}^T \mathbf{Q}^{-1}(x) \qquad (7.69)$$

The effect of $\mathbf{u}^T$ here is to sum the rows of $\mathbf{Q}^{-1}$. The multiplier λ is found by substituting back into the unit area constraint, $\Sigma\ D_j\ (x) = 1$, or

$$\mathbf{D}^T\mathbf{u} = 1 \tag{7.70}$$

Therefore

$$1 = -\lambda\ \mathbf{u}^T\mathbf{Q}^{-1}(x)\ \mathbf{u} \tag{7.71}$$

and

$$\mathbf{D}^T(x) = \mathbf{u}^T\mathbf{Q}^{-1}/\mathbf{u}^T\mathbf{Q}^{-1}\ \mathbf{u} \tag{7.72}$$

The effect of finding the best resolution without regard to noise is of course to produce considerable noise amplification. As before the noise variances in the estimate of $B_s(x)$ must be $\mathbf{D}^T\mathbf{S}_\epsilon\mathbf{D}$. If we minimize a weighted sum of spread and noise variance:

$$\frac{\partial}{\partial\mathbf{D}}\ \{\ \mathbf{D}^T\mathbf{Q}\,\mathbf{D} + \lambda\ \mathbf{D}\,\mathbf{u} + \mu\ \mathbf{D}^T\mathbf{S}_\epsilon\ \mathbf{D}\ \} = 0 \tag{7.73}$$

in order to find $\mathbf{D}$, then the solution noise variance can be controlled, but at the expense of spread. The algebra is only changed by replacing $\mathbf{Q}$ by $\mathbf{Q} + \mu\mathbf{S}_\epsilon$, so the solution is

$$\mathbf{D}^T = \mathbf{u}^T(\mathbf{Q} + \mu\ \mathbf{S}_\epsilon)^{-1}/\mathbf{u}^T(\mathbf{Q} + \mu\ \mathbf{S}_\epsilon)^{-1}\ \mathbf{u}\ . \tag{7.74}$$

We may regard μ as a 'trade-off' parameter, controlling the balance between noise and resolution. As $\mu\rightarrow0$ we get the best resolution but poor noise. As $\mu\rightarrow\infty$ we get poor resolution but best noise. This trade off between resolution and noise is physically very reasonable, and the value of μ must be chosen according to the application.

Fig. 7.6 illustrates the trade-off between noise and spread that is obtained with the weighting functions of Fig. 7.1 for a centre height of 5 scale heights. A basic noise level of $\sigma^2 = 0.1$ or $\sigma = 0.316$ has been assumed. For $\mu = 0$, the best resolution is about 1.15 scale heights with a noise of 0.72, and for $\mu = \infty$, noise is about 0.1, and spread is 4.8 scale heights. Fig. 7.7 shows the corresponding range of averaging kernels, the narrowest one being for $\mu = 0$, and the broadest is for $\mu = \infty$.

Note that the Backus-Gilbert method can be applied to cases where there is *a priori* information, simply because the *a priori* information can be treated as virtual measurements (Rodgers, 1976a).

7.7   Nonlinear problems

The general remote sounding retrieval problem is nonlinear. It is only by making simplifying assumptions that a linear problem can be constructed. The main sources of nonlinearity in the equations are:

> temperature dependence of the atmospheric transmittance;
> wavenumber dependence of the Planck function across a spectral band;
> wavenumber dependence of the Planck function between spectral bands;
> the dependence of transmittance on absorber concentration when sounding for composition;
> clouds;
> nonlinear constraints.

The first two of these usually lead to relatively small nonlinearities, whilst the rest may cause large nonlinearities.

Fig. 7.6 Trade off between noise and vertical resolutions for the idealized set of weighting functions (Fig. 7.1) at a height of $z = 5.0$.

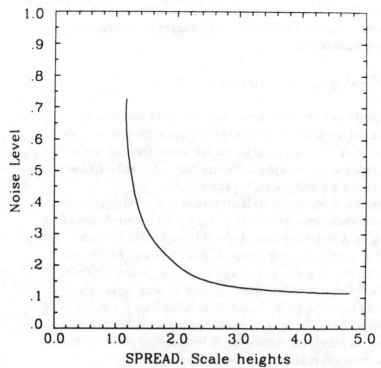

Formally the nonlinear problem can be written as a generalization
of the linear problem. We wish to know the value of an unknown vector
x. We can measure a vector **y**, which is related to **x** in a known way:

$$y_1 = F_1(x) \qquad (7.75)$$

with an error covariance S_1. We may have *a priori* information y_2 about **x**,
which is similarly a known function of **x**:

$$y_2 = F_2(x) \qquad (7.76)$$

with error covariance S_2. The process of solution is in principle the same as for
linear problems, although it may not be possible to find an explicit algebraic

Fig. 7.7 Averaging kernels at $z = 0.5$ for $\mu = 0, 1, 4, 40, 400$ and ∞, with $\mu = 0$ being the
narrowest. These are normalized to unit areas and may be directly compared with the
original weighting functions in Fig. 7.1.

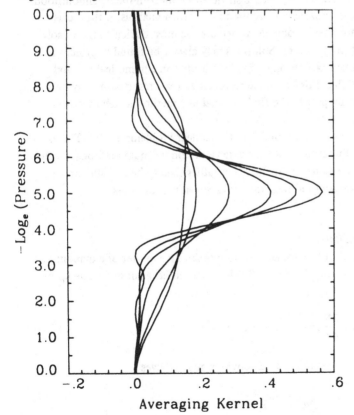

form for the best estimate $\hat{x}$. We must first ensure that there are enough measurements and constraints to determine $\hat{x}$ uniquely (i.e., with finite covariance), then set up a set of equations which $\hat{x}$ must satisfy, such as the maximum likelihood condition that $\hat{x}$ minimizes

$$\sum_i (y_i - F_i(x))^T S_i^{-1} (y_i - F_i(x)). \tag{7.77}$$

The solution of this kind of equation is usually the most difficult part of a nonlinear estimate. When the solution has been found, its error bounds must be characterized by finding its error covariance matrix.

The degree of nonlinearity in a problem can be classified in terms of the approximations that may be used to solve it. Problems which are slightly nonlinear are best solved by some form of Newtonian iteration. That is, linearize the problems and use a linear solution method, iterating as required. A second class of moderately nonlinear problems are those which need other methods to find the solution efficiently, but are sufficiently linear in the neighbourhood of the solution for linearization to be used in the error analysis. A third class is that of grossly nonlinear problems, which are the most difficult kind to solve or to understand their solutions. Solutions may always be found in principle simply by minimizing expression (7.77) by a brute-force numerical method (e.g. Barnett 1969), but this should be regarded as a last resort, as in any particular case it is usually possible to find some *ad hoc* method which exploits the algebraic form.

The correct optimum solution of nonlinear problems is hard. This is exemplified in the literature by a large variety of nonoptimum methods which use constraints designed to aid the iteration rather than to be realistic, or which converge to exact (and therefore noise-sensitive) solutions.

7.7.1   Newtonian iteration

The method of Newtonian iteration is simply a matter of expanding the direct model (equation 7.75) as a Taylor series about a guessed value x_n of the solution

$$y = F(x_n) + \frac{\partial F}{\partial x} (x - x_n) + O(x - x_n)^2. \tag{7.78}$$

Redefining some of the symbols in an obvious manner gives

$$y = y_n + K_n (x - x_n) + O(x - x_n)^2 \tag{7.79}$$

The higher order terms are ignored, leaving a set of underconstrained linear equations for x, which are solved in the usual way, using an *a priori* constraint x_0 with covariance S_x. The equations are now:

$$K_n x = y - y_n + K_n x_n, \text{ covariance } S_\epsilon \tag{7.80}$$

$$x = x_0, \qquad\qquad \text{covariance } S_x \tag{7.81}$$

Thus the solution is

$$x_{n+1} = (S_x^{-1} + K_n^T S_\epsilon^{-1} K_n)^{-1} \left\{ K_n^T S_\epsilon^{-1} (y - y_n + K_n x_n) + S_x^{-1} x_0 \right\} \tag{7.82}$$

using equation (7.17). Here $\hat{x}_{n+1}$ is written in place of $\hat{x}$ because this is an iteration equation, and $x_n \to \hat{x}$ as $n \to \infty$. This may be rearranged to give $\hat{x}_{n+1}$ as a departure from either x_n or from x_0

$$x_{n+1} = x_n + (S_x^{-1} + K_n^T S_\epsilon^{-1} K_n)^{-1} \left\{ K_n^T S_\epsilon^{-1} (y - y_n) + S_x^{-1} (x_0 - x_n) \right\} \tag{7.83}$$

$$x_{n+1} = x_0 + (S_x^{-1} + K_n^T S_\epsilon^{-1} K_n)^{-1} K^T S_\epsilon^{-1} \left\{ (y - y_n) - K_n (x_0 - x_n) \right\} \tag{7.84}$$

However the computationally efficient form can only be derived from the second of these:

$$x_{n+1} = x_0 + S_x K_n^T (K_n S_x K_n^T + S_\epsilon)^{-1} (y - y_n - K_n (x_0 - x_n)) . \tag{7.85}$$

This process has second-order convergence, because it gives the solution in one step in the linear case, so the value of $x_{n+1} - x_n$ may be used as a convergence criterion. The covariance of the solution is of course

$$\hat{S} = (S_x^{-1} + \hat{K}^T S_\epsilon^{-1} \hat{K})^{-1} \tag{7.86}$$

or its equivalent according to equation (7.24), where $\hat{K}$ is K evaluated at $\hat{x}$.

The main drawback of Newtonian iteration is that the Fréchet derivative $K = \partial F/\partial x$ must be evaluated at each stage, in contrast to the linear case where almost everything can be precomputed, and the solution reduces to a matrix multiplication. However in some nearly linear cases a constant can be used for K. Convergence will be obtained but the solution will not be optimum. The non-optimum solution can then be used as a new starting point where a new value of K is computed. This approach will require more iterations, but fewer evaluations of K_n.

A trap that some authors have fallen into is to iterate the linearized equations using the *n*th estimate x_n as the *a priori* information for the (*n*+1)th stage, i.e., replacing x_0 by x_n in the above equations. An example of this is

Smith, Woolf and Fleming (1972) who have used Twomey's method this way. The effect is that the original *a priori* contraint is lost, and the solution converges to the undesirable 'exact' solution.

Surmont and Chen (1973) have used a Newtonian iteration without apparent constraint, in such a way that it converges onto an exact solution. They found that, in the absence of experimental noise, a solution was obtainable for 7 spectral intervals, but was numerically unstable for 12 spectral intervals. Since they did not include experimental error, the numerical instability was presumably due to computational truncation error being amplified by an inherently unstable method.

7.7.2   *Moderately nonlinear problems*

If a problem is too nonlinear for a linearization method to be efficient, then there are two courses of action available. The estimation equation can be solved by any of the general purpose routines to be found in computer program libraries, either minimizing expression (7.77) directly or solving the nonlinear equations obtained by differentiation of (7.77). This process certainly works, but is likely to be inefficient and there is always the possibility of a non-unique solution. Provided the equations are sufficiently linear in the neighbourhood of the solution, then the solution covariance is still given by equation (7.86)

A more practical approach is to make use of the nature of the equations to reduce the amount of work to be done, and to use some *ad hoc* method to find a solution. If necessary the instrument should be designed with the solution of the retrieval problem in mind. *Ad hoc* solutions are usually nonoptimum, but they may lie within the population defined by the optimum solution and its covariance. If this is so, then a final stage of linearization will give an optimum solution.

There are many *ad hoc* approaches to be found in the literature, some of which are listed here: Barnett (1969), Chahine (1968, 1970, 1972, 1974), Chow (1974), Conrath (1969), Gille and House (1971), King (1964), Scott and Chedin (1971), Smith (1967, 1968, 1970). Most of these authors seem not to realize that their solution is nonoptimum, and that they have found only one of a population of equally valid solutions. Consequently, the final step of linearization is not usually taken. In order to keep this chapter within reasonable bounds, we will only describe a few of the nonlinear *ad hoc* methods.

7.7.3   *Chahine's method*

Chahine's method relies on the weighting functions having well defined peaks, so that the radiance measured is closely related to the temperature in the region of the peak. The atmospheric temperature profile is represented by linear interpolation between temperatures at fixed levels defined by the peaks of the weighting functions. (A variation due to Smith (1970) has the temperature profile represented by a linear combination of weighting functions). At any stage of the iteration, radiances are computed for the current estimate of the temperature profile, and the profile is then adjusted by the iteration equation:

$$B(\theta_j^{n+1}, \nu_j) = B(\theta_j^n, \nu_j) I_j^{\text{obs}}/I_j^{\text{calc}}(\theta) \qquad (7.87)$$

where θ_j^n is the temperature at level j (corresponding to weighting function j) at stage n of the iteration, $B(\theta, \nu)$ is the Planck function, I_j^{obs} is the observed radiance, and $I_j^{\text{calc}}(\theta)$ is the calculated radiance for the profile defined by $\theta = (\theta_1, \theta_2 \ldots \theta_M)$. The value of θ_j^{n+1} is found by using the inverse Planck function. It can be seen intuitively that if the weighting functions are reasonably well peaked and separated then the effect of this iteration is to produce a profile such that

$$\lim_{n \to \infty} I_j^{\text{calc}}(\theta^n) = I_j^{\text{obs}}. \qquad (7.88)$$

If the iteration is allowed to proceed indefinitely, then it will converge to an undesirable exact solution. However a convergence analysis shows that broad-scale features converge rapidly, and fine-scale features converge slowly, so that a qualitatively 'reasonable' solution may be found if the iteration is stopped at the right point. In the case of a linear problem, Chahine's method will eventually converge to equation (7.6) for the appropriate form of the representation $W_j(z)$, and Smith's version will converge to equation (7.8). The main advantage of Chahine's method is that it is easy to understand and to program. However, the error analysis is difficult, and we do not know which of the infinite set of possible solutions will be produced. In a fairly strongly nonlinear problem it may converge faster than a linearization method, so that in this case an effective combination might be to use a Chahine solution as a first guess for an optimum linearization method.

7.7.4 Adjacent fields of view

Clouds are probably the most important source of nonlinearity for tropospheric soundings. Techniques exploiting the properties of adjacent fields of view, in which the thermal structures are nearly identical but the cloud amount may vary, have been developed to tackle this problem. These are considered in more detail in § 8.3.

7.7.5   Limb sounding

An important method of improving the vertical resolution of remote
sounding is to use limb sounding. Here, one spectral interval and a narrow
field of view is used, directed at the limb (Fig. 5.1) so that a large proportion of
the radiation reaching the instrument originates near the tangent height, resulting in narrow weighting functions. A typical set is shown in Fig. 5.5. Nonlinearities arise from three sources in this case:

temperature dependence of transmittance

the hydrostatic equation gives a variable relationship between tangent
height and pressure;

spacecraft attitude is often not well enough known, so that absolute
tangent height must be derived from the measurements;

If temperature dependence of transmittance were the only source of
nonlinearity, we would have an ideal case for sequential estimation using
equations (7.53), introducing the measurements in order, starting at the top,
and calculating each weighting function as it is needed, based on the retrieval
temperatures above the level being considered. This has been described as the
'onion-peeling' approach in its nonoptimal form (e.g. Russell and Drayson
1972).

If the spacecraft attitude is known, then the sequential estimation
method could be iterated to allow for the effect of redistribution of absorber
amount with height due to changing the temperature profile, but it is not
obvious that the process will converge, or converge rapidly enough.

However, the spacecraft attitude is not usually known well enough,
so that it becomes necessary to design the experiment to measure attitude, or
to find its own reference level. This has been done in the case of the LRIR
instrument (Ch. 6) by measuring limb radiance in two spectral intervals, a
narrow band and broad band. To a first approximation the ratio of the two
signals is independent of the temperature profile, and is equal to the ratio of the
absorptivities of the emitting gas (carbon dioxide), which is a known function
of atmospheric pressure. Thus the direction of a known pressure level may be
found, and may be improved by iteration (Gille and House, 1971).

8

REMOTE TEMPERATURE SOUNDING
FOR WEATHER FORECASTING

8.1 Introduction

In order to forecast the weather, the meteorologist has access to a
wide range of measurements of the atmosphere. These include a network of
surface observations of temperature, wind, pressure, humidity and cloudiness,
and radiosonde measurements of temperature, pressure, humidity and wind up to
an altitude of 20 or 30 km. Reports from ships and aircraft are also available. A
global telecommunications system exists whereby this data is distributed to forecasting centres in a timely fashion. Satellites have made images available in both
the visible and the infrared allowing cloud systems to be seen, and measurements of thermal emission which allow the distribution of temperature to be
derived.

The network of radiosonde and surface observations is very unevenly
distributed, being concentrated in areas of relatively high technological development, and almost absent over the oceans. The southern hemisphere is particularly
poorly covered, due to the relatively small area of land.

Satellites make measurements continuously and with a much more uniform spatial coverage, thus providing a uniform data set. For convenience of
analysis, surface-based measurements are made at the so-called 'synoptic times'
of noon and midnight G.M.T. The continuous or asynoptic nature of satellite
measurements, therefore, causes practical problems of analysis. Furthermore a
temperature sounder does not measure temperature directly as a function of
height, rather it measures a weighted mean of Planck function over some range
of heights. It has been found convenient to use these radiances to retrieve temperature profiles which are then used in the same way as radiosonde profiles in
the analysis.

The most important region for forecasting is the troposphere, although
the lower stratosphere is represented in most forecasting models. In the lower
atmosphere the main problem for remote sounding is the presence of clouds

which are ill-defined in shape and more or less opaque. They shield the atmosphere below from the line of sight of the sounding instrument. Techniques have been developed which can recover the radiance which would be seen in the absence of cloud, at least for some partly cloudy areas. The resulting 'clear column radiances' are then used to retrieve temperature profiles which are combined with all the other available information about the state of the atmosphere, thus determining an initial value of the distributions of meteorological parameters which may be used as a boundary condition for an integration of the equations of motion to produce a forecast.

8.2   Requirements of meteorology

Bengtsson (1979) has reviewed the problems of using satellite information in numerical weather prediction. He gives the accuracy and resolution requirements for large-scale prediction shown in Table 8.1.

Numerical prediction involves integrating the simultaneous nonlinear partial differential equations which describe the motion of the atmosphere from an initial state, using a large-scale grid. This initial state must be known for an area which increases with the length of the forecast period, the linear dimension increasing at a rate which is related to the group velocity of certain wave motions. This velocity is typically $40\,\mathrm{ms}^{-1}$, and consequently initial information is needed from a hemispheric domain if the forecast period is four or five days, and from the whole globe if we want to predict more than a week ahead.

A surface-based global network to provide information satisfying the requirements of the table would need a few thousand stations launching radiosondes four times a day, uniformly spaced over the globe. Two-thirds of them would be weather ships. Such a network would of course be prohibitively expensive. The existing radiosonde network is more than adequate over some populated areas, but is very inadequate over the oceans, and especially in the southern hemisphere.

Table 8.1. *Accuracies and resolution for parameters needed for large-scale numerical weather prediction according to Bengtsson (1979)*

Parameter	Accuracy
surface pressure	1 mbar
horizontal wind	$3\,\mathrm{ms}^{-1}$
temperature	1 K
relative humidity	5 %
horizontal resolution	500 km
vertical resolution	1 km
time resolution	6 hr

Satellite sounding would appear to be the only solution because it can easily provide the horizontal and time resolution at a much lower cost than a radiosonde network, but unfortunately it cannot satisfy the other requirements given in the table. It has not yet proved possible to measure surface pressure remotely, although some proposals for doing this have been made (§ 12.1). At present wind can only be measured directly by tracking clouds in satellite pictures. Whilst this is of considerable value, it can only be carried out satisfactorily when there are easily identifiable individual clouds present, and even then there are problems of the altitude being measured, and the identity and stability of the individual cloud as an inert tracer.

As shown in Ch. 7, vertical resolution and accuracy of temperature profile retrievals are interrelated. It is possible to measure temperature to a precision considerably better than 1 K at the basic vertical resolution of the measurement, typically $\sim 10\,km$, but if attempts are made to improve the resolution by profile retrieval, then a loss of accuracy must occur. Satellite measurements are likely to have systematic errors which are greater than 1 K, but this is comparable with the systematic differences between different types of radiosonde sensor, and may be removed by intercomparison.

It should be mentioned in passing that radiosondes suffer from a resolution problem too, but in this case it is too high. Their observations are contaminated by subgrid scale fluctuations which are not present explicitly in the numerical model, and hence must be regarded as noise.

8.3   Determination of clear column radiances

In order to make proper use of tropospheric soundings it is necessary to deal with the problem of cloud contamination. The simplest approach is to use microwave channels for which clouds are nearly transparent. This is done on the operational temperature sounder on the TIROS-N Series of spacecraft, but the vertical resolution is relatively poor, and in clear areas more information can be obtained from infrared channels. It is, therefore, important to obtain as much information as possible from the infrared channels in partly cloudy areas. The technique used operationally exploits the properties of adjacent fields of view, as has been described by Smith (1968). Operational sounding instruments have been designed with this method in mind by scanning in such a way that there are many adjacent fields of view.

The principle on which the method is based is that clouds have a considerable amount of spatial structure in the horizontal, in relation to other variables in the atmosphere, so that in two sufficiently small adjacent fields of view the temperature profile is essentially the same, and the cloud height (or distribution in height) is likely to be the same, but the cloud amount may be quite different.

In the simplest case of one layer of cloud at height c, the radiance is given by

$$L(\nu) = \int_0^\infty B(\nu, z) \frac{dT(\nu, z)}{dz} \, dz + n \int_0^c \left[B(\nu, c) - B(\nu, z) \right] \frac{dT(\nu, z)}{dz} \, dz \qquad (8.1)$$

where n is the cloud amount. This may be written more simply as

$$L = L_0 + n \, \Delta L(c) \qquad (8.2)$$

where L_0 is the 'clear column radiance' and $\Delta L(c)$ is the correction to L_0 for a complete cloud cover at height c.

We can make use of the form of this equation to eliminate $\Delta L(c)$ and solve for L_0 in several ways. For example, if two adjacent fields of view have measured radiances L_1 and L_2 and cloud amounts n_1 and n_2, then

$$L_1 = L_0 + n_1 \Delta L(c) \qquad L_2 = L_0 + n_2 \Delta L(c) \qquad (8.3)$$

When $\Delta L(c)$ is eliminated

$$L_0 = \frac{n_2 L_1 - n_1 L_2}{n_2 - n_1} = \frac{L_1 - n^* L_2}{1 - n^*} \qquad (8.4)$$

where $n^* = n_1/n_2$. If there is an independent measurement of surface temperature, together with measurements of L_1 and L_2 in a spectral window, then n^* can be determined and used to solve for L_0 in other spectral intervals.

Another approach, due to Smith and Woolf (1976) reduces equation (8.4) to a nearly linear form,

$$L_1 = (1 - n^*) L_0 + n^* L_2$$

If $(1 - n^*)L_0$ and n^* are regarded as the unknowns, then this is a linear equation. Unfortunately, this leads to more unknowns than equations, so that *a priori* constraints are required. Smith and Woolf apply these by representing the vector of measurements (19 in their case) by an empirical orthogonal function expansion with 10 terms, and then solve the resulting over-constrained linear equations by least squares. This would appear to be a linear solution to a non-linear estimation problem. Unfortunately, it is not, because the least-squares solution is not optimum. The optimum solution leads to a weighted least-squares solution, where the weights depend on the solution. The problem has been reduced to a linear form, but the *a priori* constraints become nonlinear. Nevertheless the method works well and produces reasonable results. In the TIROS-N global sounding operation, soundings are derived from radiances for a 9 x 6 array of HIRS fields of view and spatially interpolated radiances from the microwave sounding unit. The clear column radiance determinations for partly cloudy measurements are combined with cloud-free measurements to form a spatial average for an area of about 250 km across, after appropriate quality checks.

8.4   Temperature retrievals

Once clear column radiances are available, temperature retrievals are straightforward in principle, and any of the techniques described in Ch. 7 may be used. The TIROS-N operational system uses the method of multiple regression where the coefficients are obtained from a contemporary colocated sample of measured clear column radiances and radiosonde observations. The samples are grouped into latitude zones. Eigenvector (empirical orthogonal function) regression coefficients are derived for levels below the 100 mbar level, while ordinary linear regression is used for stratospheric levels. In order to avoid discontinuities along the latitudes used to group the statistical data, the regression coefficients are interpolated to specific sounding conditions using a microwave brightness temperature as the interpolation variable. In the case of extensive overcast cloud conditions, a set of microwave and stratospheric channel infrared temperature profile regression relations are used for the profile determination. The regression technique has the advantage that it is not necessary to know the spectral response or even the calibration of the instrument to high accuracy, all that is required is that the response be adequately linear within the possible range of variation. More physical methods, such as those that need to know the equation of transfer and the weighting functions can allow for nonlinearities and dependencies on parameters rather than temperature, but are sensitive to errors of calibration and the physical definition of the instrument.

8.5   Objective analysis methods

Conceptually, the first stage in producing a weather prediction is the 'analysis' or determination of the initial state of the atmosphere on the spatial grid that will be used for the numerical integration. This involves combining together all the information available at the analysis time to produce a best estimate, and as such it is a retrieval problem of the same kind as discussed in Ch. 7, but of enormously greater proportions. In practice, objective analysis is intimately bound up with the forecasting process because a forecast will be used as a 'first guess field' (i.e. *a priori* information) which is updated by the measurements. In the past this updating took place for synoptic times when most of the measurements were made, but more recently, with more and more asynoptic observations such as satellite data, it has become more of a continuous process, described as 'four-dimensional assimilation'.

The *successive correction method* is originally due to Bergthorssen and Döös (1955), and has been further developed by Cressman (1959) amongst others. Basically the method consists of updating a first guess field, usually a forecast, with synoptic time observations by computing a weighted mean value of the measurements at stations surrounding a grid point, the closest station having the highest weight, according to a predetermined weighting function. Thus for a quantity q_i to be analysed at grid point i, the estimated value would be

$$\hat{q}_i = q_i^0 + \sum_k p_k \, (q_k^m - q_k^0)/\sum_k p_k \tag{8.5}$$

where q^0 is the *a priori* value, q^m is the measured value, and p_k is the weighting function which might be of a form such as $a \exp(-br^2)$, $a/(b + cr^4)$ or $(a - r^2)/(a + r^2)$ where a, b and c are empirical constants and r is the distance between the measurement and the grid point. Operationally one usually carries out several scans through the data; the result of one scan being used as a first guess for the next. The form of p_k is changed on each scan so that observations farther from the grid point have less weight on successive scans.

Optimum interpolation (Gandin 1963), uses the methods of statistical estimation theory introduced in Ch. 7 as a profile retrieval method. The statistical nature of the first guess field is known or approximated, typically as a covariance with a form such as $a \exp(-br^2)$, i.e. homogeneous and isotropic. The weights p_k in the update equation can be calculated from the statistics of the measurement and the first guess field using minimum variance estimators. The advantage is, course, optimality, insofar as the assumed statistics are correct, but the process uses rather more computer time. It is straightforward to include such things as correlated measurements, meteorological constraints, and the simultaneous analysis of all the relevant meteorological parameters.

Flattery (1970) has developed a technique of surface fitting using Hough functions as basis functions in the horizontal, and a set of empirical functions in the vertical. Geopotential (z) and wind (u, v) are jointly analysed by least-squares fitting, where a sum of the form

$$S = \sum \{ \xi(z^m - z^0)^2 + (u^m - u^0)^2 + (v^m - v^0)^2 \} \tag{8.6}$$

is minimized with respect to the fitting coefficients. The weighting factor ξ allows relative weighting of measurements of geopotential against measurements of wind. The Hough functions are a natural set of functions to use for the atmosphere, because they are eigenfunctions of an approximation to the equations of motion of a fluid on a sphere.

All the above methods are designed for use at synoptic times, and need modification for the assimilation of asynoptic data. Two basically different approaches have been used for this purpose, namely *intermittent data assimilation* or *direct insertion*. In the first of these, asynoptic data is grouped at sets of discrete times, generally at 6 hour intervals, and a three-dimensional assimilation method is used. This ensures that data is used not more than 3 hours away from its true time, but nevertheless significant changes may have taken place. In the direct insertion approach, a forecasting model is run, and data is inserted in the field at the appropriate time and place, with some spatial weighting. It is assumed that the model can absorb this single piece of information in a dynamically consistent way. This too can have problems, because shocks are introduced into the model in a regular way in the case of satellite data. It is important to minimize the effect of such shocks, as they may propagate as unwanted waves in the forecasting model.

McPherson & Kistler (1974) have shown that both wind and mass field data must be inserted simultaneously into the model in order to guarantee a satisfactory data assimilation. The use of a statistical consistent analysis procedure such as three-dimensional optimum interpolation or multivariate scheme (Rutherford, 1976) has also been found to be very successful in this respect. Lorenc *et al.* (1977) have analysed all available data in a 6 hour interval using such a scheme, incorporating the actual statistical structure of observational error, including the systematic errors of satellite retrievals. Even a very accurate statistical analysis does not guarantee a noise-free initial data set, and some sort of *initialization* is necessary. This can be done by using a frequency selective damping scheme, or a time filter, or using forward–backward time integration of the model. Recently a new, very powerful method, nonlinear normal mode initialization, has been developed (Machenhauer 1977, Baer 1977) which efficiently eliminates all gravity waves from the initial state.

8.6 Impact of satellite soundings on forecasts

Evaluations of the impact of satellite soundings on forecasts have been carried out by a number of groups using a variety of analysis and forecasting methods. Table 8.2, taken from Ohring (1979), shows a comparison of impact studies in terms of S1 skill score. The S1 score is defined as

$$100 \, \Sigma \, |\epsilon_G| / \Sigma \, |G|$$

where ϵ_G is the error of the forecast height difference between grid points, and G is the actual or forecast height difference, whichever is the larger. Thus lower scores correspond to greater skill.

The largest improvement is obtained in the southern hemisphere, as would be expected in view of the relative lack of conventional data there. A case taken

Table 8.2 *Impact of satellite soundings on S1 scores of 48 H geopotential height forecasts. (Positive impact represents reduction in S1, improvements in forecast.)*

Source	Verification area	Season	Data	Number of forecasts	NOSAT	SAT	IMPACT
Desmarais et al. (1978)	N. America	Summer	V + N	10	46.0	46.1	-0.1
Halem et al. (1978)	N. America	Winter	V + N	15	34.8	34.3	0.5
Bonner et al. (1976)	N. America	Winter	V + N	11	39.6	37.7	1.9
Kelly (1977)*	Australia	Spring	V	9	44.2	45.6	1.4
Kelly et al. (1978)**	Australia	Winter	V	9	42.0	40.0	2.0
	Australia	Winter	N	28	42.2	37.4	4.8

V: VTPR soundings, N: Nimbus 6 soundings
*: 36 h forecasts
**: 24 h forecasts

from Kelly *et al.* (1978) is shown in Figs. 8.1 and 8.2 to illustrate the effect on a forecast of the difference in skill that they found. The figures show the initial surface pressure analysis (8.1*a*), 24 hour prognoses with (8.2*b*) and without (8.2*a*) Nimbus 6 data, and the verifying surface analysis (8.1*b*). The Nimbus prognosis maintains the intensity of the low pressure system in the Indian Ocean coast, and moves it eastward across the coast, although not quite so far as it did in fact move. The control prognosis keeps the low off the coast, and increases its central pressure by 5 mbar. Both prognoses over-intensify the anticyclone over Australia and fail to move it east, but this is less severe in the Nimbus prognosis.

Two of the studies in the table (Desmarais *et al.* (1978) winter, and Halem *et al.* (1978)) were carried out on the same data set as part of the data

Fig. 8.1 Surface pressure analyses for the Australian region, including satellite data. (*a*) 23 August 1975 (*b*) 24 August 1975.

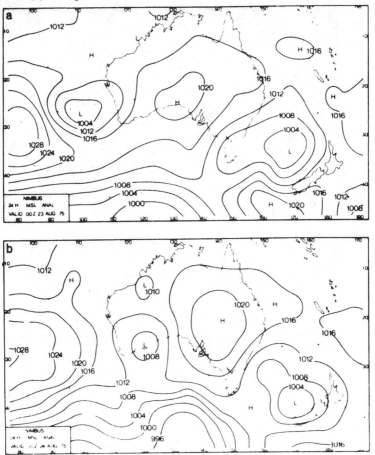

systems test (DST) which was set up in preparation for the first GARP global experiment. The differences between their conclusions illustrate the extreme complexity of this problem.

The first GARP global experiment (FGGE) has resulted in a very comprehensive data set becoming available, allowing much more detailed studies of satellite impact on weather forecasting to be carried out. A programme has been set up under the auspices of the GARP Working Group on Numerical Experimentation to examine the impact of various observing systems in defining the synoptic and larger scale features of the general atmospheric circulation used as an initial state for numerical weather prediction. Their report, following a

Fig. 8.2 24-hour prognoses for 24 August 1975 (*a*) using no satellite data in the initial field (*b*) including satellite data in the initial field.

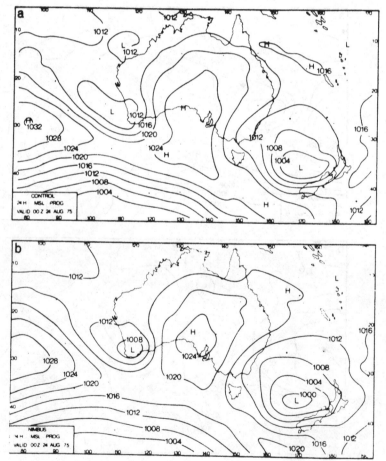

study conference at Exeter University (Gilchrist 1982) shows that satellite data does indeed have a considerable effect on the predictability of the weather.

In an experiment carried out at the European Centre for Medium-range Weather Forecasting (ECMWF), forecasts were compared in which (a) all the available FGGE data was used and (b) all the satellite data (both temperatures and cloud-drift winds) were omitted. The average period of 'useful predictability' (defined as the number of days for the 1000 to 200 mb height anomaly correlation to drop below 60%) was reduced from 5½ to 4½ days in the northern hemisphere, and from about 5 days to less than 3 days in the southern. There were substantial phase differences in major trough-ridge systems after a few days, and they increased with time. This impact of satellite data was therefore very clear.

In a similar exercise carried out at the Goddard Laboratory for Atmospheric Science (GLAS), using 300 mb heights, the comparisons demonstrated that large 6h forecast errors downstream of data-sparse regions were reduced by as much as 50% when satellite observations were available to the analysis scheme. Comparison of skill scores on a geographical basis indicated that there was a large positive impact over Australia and South America. Over North America and Europe, the impact was smaller but still positive, and it became more significant after a period of 3–5 days.

In studies at GLAS involving the impact of satellite temperatures only, the impact was examined statistically over the Australian region, where it was shown that during the first special observing period (SOP-1) the forecasts were degraded as a result of omitting satellite temperatures. At 500 mb, skill scores were less by 1–3% at 24h, 4–6% at 48h and 8–10% at 72h. During SOP-2, the degradation was less, presumably indicating a relatively greater impact of other FGGE systems. Over North America and Europe, the impact was smaller and more variable; it reached its highest values in forecasts at about 72h.

In two assimilation/forecast experiments at ECMWF (Bengtsson and Kallberg, 1981) using SOP-1 data, forecasting performance of a hypothetical space-based observing system consisting of surface pressure data from buoys, wind data from aircraft, cloud track winds from satellite imagery, constant-level balloon observations and satellite-derived temperatures was compared with the complete FGGE observing system. Although the predictability in the northern hemisphere was somewhat lower in the two space-based forecasts, they were still surprisingly good with an average predictive skill of about 4 days. The corresponding figure for the complete system was about 5 days. In the southern hemisphere there were very small differences, both having a predictive skill of about 5 days. Synoptic developments at southern hemisphere middle and high latitudes were very similar.

Betout (see Gilchrist 1982) has experimented with the direct assimilation of radiance data into an objective analysis scheme. In principle this is desirable because the unnecessary intermediate step of a temperature profile retrieval is avoided. A retrieval profile is necessarily smoother in the vertical than the actual profile, and the operational retrievals appear systematically to underestimate horizontal gradients. The objective analysis scheme used optimum interpolation, so it was straightforward to assimilate radiances directly. The results of this simple experiment using a forecast over North America and Northern Pacific Ocean were disappointing; only a small impact was noticed in the analyses, and no impact was seen in the forecasts. However, it may prove that further studies using better representations of the satellite radiance error covariance field will show the expected improvement.

Satellite radiances can be measured, in principle, to a very high precision (typically better than 0.2 K in equivalent temperature). Much of the error in retrievals is systematic or is due to the treatment of clouds. Although the observations have poor vertical resolution, they have very high horizontal resolution, providing a wealth of detailed information which has been successfully employed in research studies and in mesoscale forecasting (Smith *et al.* 1979).

9

STRATOSPHERIC AND MESOSPHERIC
TEMPERATURE MEASUREMENTS

9.1   Introduction

Until remote-sounding data for the stratosphere and mesosphere (often called the middle atmosphere) became available, the only routine information was from radiosondes for the low stratosphere and from rocketsondes for the whole of the region. Needless to say, rocket data is very sparse, and it is difficult to obtain a complete global picture of the structure and dynamics of the middle atmosphere from this source alone. With the launch of the first sounders, it was suddenly possible to map the temperature distribution in three dimensions, and to follow its changes day by day. The intimate relationship between the temperature field and the motion field which is a consequence of the equations of motion makes it possible to carry out a whole range of studies on the dynamics of the upper atmosphere.

The use of satellite sounding has the major advantage of making global and fairly uniformly distributed measurements with a single instrument, thus eliminating the need for cross calibration between many different types of *in situ* instruments. One instrument can supply a global data set of uniform quality.

Most of the instruments described in Ch. 6 have channels which sound parts of the middle atmosphere. Some have been designed specifically with this part of the atmosphere in mind, namely the SCR on Nimbus 4 and 5, the LRIR and PMR on Nimbus 6, the SSU on TIROS N and the NOAA series, and LIMS and SAMS on Nimbus 7. These instruments, together with the upper channels of the tropospheric sounders have provided a wealth of information about the stratosphere and mesosphere since 1970.

9.2   Annual and semiannual variation

Climatologies of the middle atmosphere have been built up from rocket soundings and other surface-based information, and the set due to Groves (1970) has been adopted as the CIRA (1972) standard atmosphere. Soundings from satellites give us considerably more information, both about the zonal mean distribution averaged over months or seasons, and about the nature of departures from this mean in time and longitude. Labitzke and Barnett (1979) have constructed zonal mean temperature cross sections based on SCR and PMR radiance measurements for the height range 10–80 km and for the months of October, January, April and July. These cross sections are shown in Figs. 9.1 to 9.4. Also shown as dashed lines, are differences from the CIRA 72 climatology for the northern hemisphere, assuming that the southern hemisphere is the same as the northern for the same season.

The main features are qualitatively similar to the CIRA atmosphere, namely:

(1) Relatively small variations through the year in equatorial regions, mainly because solar heating varies little here.

(2) The stratosphere and stratopause are colder over the winter pole than over the summer pole, as one might expect.

(3) The mesopause region is warmer over the winter pole than over the summer pole, contrary to expectations.

Fig. 9.1 Zonal mean temperature cross section (October).

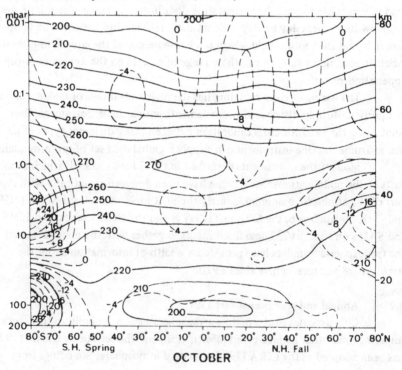

(4) Weak meridional gradients at the stratopause and above in the transitional seasons.

(5) Interhemispheric differences for corresponding seasons.

The differences from CIRA are largest over the polar regions in winter, where midwinter disturbances make it rather difficult to establish a representative climatology. They are smaller over the summer hemisphere and over tropics and subtropics at all times of year.

Studies of the annual and semiannual cycles in the climatology have been made by several authors, including Fritz and Soules (1972), Barnett (1974), Chapman and McGregor (1978) and Crane (1979b). Each study used data from a different time and a different instrument, but the findings are in general agreement. The annual cycle has its largest amplitudes at the poles near the stratopause of about 20 K in the Northern Hemisphere and about 30 K in the Southern Hemisphere. The times of maxima are close to the summer solstice. The Southern Hemisphere amplitude is larger, probably because the eccentricity of the earth's orbit around the sun causes the Southern Hemisphere winters to be colder, and summers to be warmer, than the Northern Hemisphere, and because the Northern

Fig. 9.2 Zonal mean temperature cross section (January).

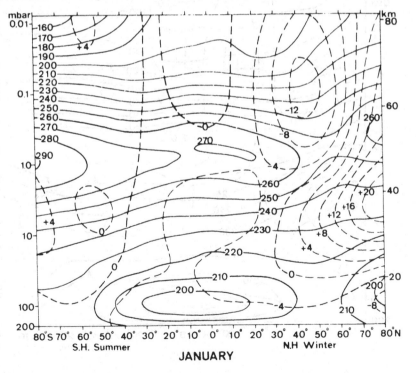

Hemisphere experiences larger stratospheric warmings. The global mean temperature for the stratosphere undergoes a substantial annual variation, apparently because of the eccentricity of the earth's orbit, which causes a 7% difference in insolation between December and June. The global mean temperature variation is in phase with this, but is of a smaller amplitude than one might expect from very simple considerations. The semiannual wave has amplitude maxima near the pole which appear to be a combination of midwinter warmings and solar heating in the summer. The amplitude would, therefore, be expected to vary from year to year. There is also a 4 K amplitude maximum in the tropical upper stratosphere which dominates the temperature variation there. Its phase is the opposite of that of the polar cycles. It is in phase with the twice yearly passage of the sun overhead, but it may also be forced dynamically from the winter hemisphere through the set of phase relationships to be discussed below.

At higher levels, in the region of the mesopause, the annual and semi-annual cycles have opposite phase to those at the stratopause in mid and high

Fig. 9.3 Zonal mean temperature cross section (April).

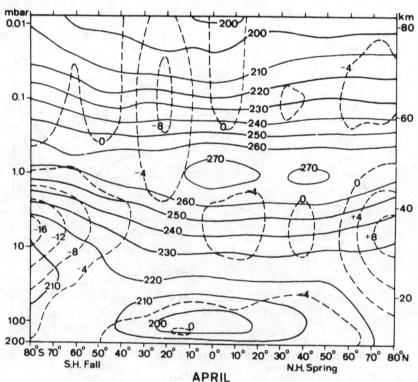

APRIL

latitudes, but are in phase at low latitudes. Thus, the semiannual cycle is in phase (maximum at equinoxes) at all latitudes, and the global mean has a strong semi-annual component, comparable in size with the annual wave amplitude of about 2 K. The annual wave in the global mean is in phase with solar forcing.

9.3 The global circulation

Once the distribution of temperature in space and time has been determined, it is possible to carry out a large range of diagnostic studies with the aid of a fairly small amount of extra information from *in situ* measurements. If the height of the 100 mbar surface, say, is available from the radiosonde network, then the heights of all the pressure surfaces within the range of the temperature sounder can be obtained by integrating the hydrostatic equation:

$$\frac{\mathrm{d}p}{\mathrm{d}z} = \frac{g}{R} \frac{p}{T} \qquad (9.1)$$

Fig. 9.4 Zonal mean temperature cross section (July).

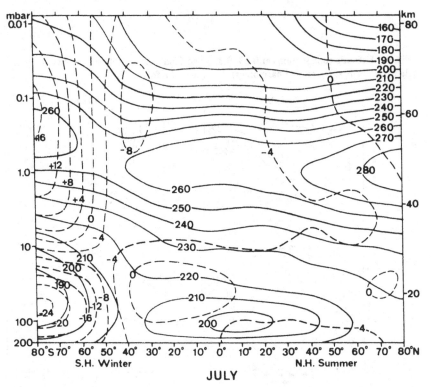

JULY

The next stage is to obtain horizontal winds by using the geostrophic approximation, which is of adequate accuracy outside tropical regions. This approximation assumes that pressure gradient forces are balanced by the Coriolis force. If an acceleration term is included to allow for the curvature of streamlines, the approximation is improved further. Once the temperature and horizontal wind are found, the vertical component can be obtained, albeit to a lower accuracy, with the aid of the momentum, continuity, and thermodynamic equations, provided the radiative cooling of the atmosphere can be calculated. The basic information is now available for the calculation of transports of heat, momentum, and energy by mean motions and by eddies. These transports are important in diagnostic studies of the global circulation and dynamics of the atmosphere.

One such study has been carried out by Crane *et al.* (1980) for the Northern hemisphere for a period of one year (1976) using radiances from the PMR, and the Berlin 30 mb height field as a base. Their latitude time cross-sections of eddy momentum flux and eddy heat flux are shown in Figs. 9.5 and 9.6. The most obvious feature of these plots is the large northward fluxes of both quantities in the winter seasons, and the hint of interannual variability in the winters in the way that January 1977 is clearly not going to be like January 1976.

Fig. 9.5 Latitude-time sections at 0.03, 0.3, 3 and 30mb of monthly mean eddy momentum flux (sum of first four zonal harmonics), $\overline{u'v'}$ (m^2 s^{-2}), for 1976.

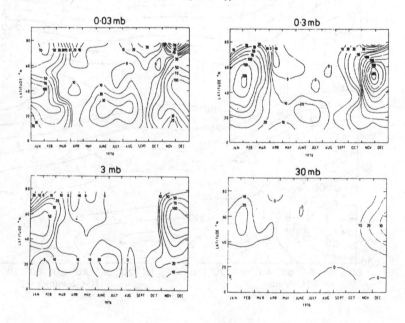

Fig. 9.6 Latitude-time sections at 0.03, 0.3, 3 and 30mb of monthly mean eddy heat flux (sum of first four zonal harmonics), $\overline{v'T'}$ (K ms^{-2}), for 1976.

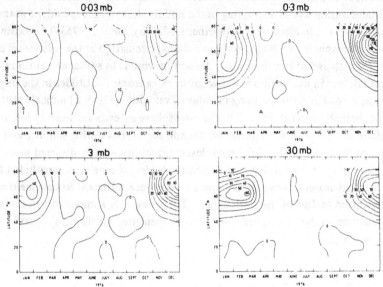

Fig. 9.7 Zonal mean vertical velocity (mms^{-1}) at 0.03, 0.3, and 30 mbar plotted against latitude for March, June, September and December 1976. Positive values denote upward motion. The bold, unjoined points shown for June are values calculated when $\overline{v'T'}$ assumed zero everywhere.

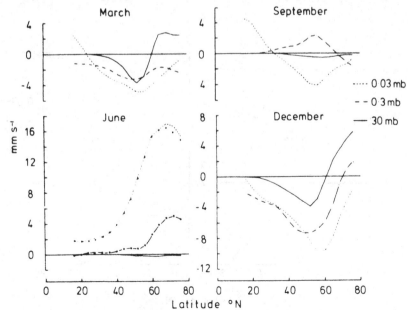

Fig. 9.7 shows the calculated vertical velocities for March, June, September and December. A high latitude cell with ascent over the pole and descent in mid latitudes during winter and spring in the middle stratosphere is evident in the 30 mbar values, in agreement with other studies by Adler (1975). The existence in the mesosphere of a global-scale meridional circulation at the solstices is confirmed, with ascent in summer and descent in winter. In spring and autumn ascent occurs in the tropics and descent in extra tropical latitudes in the upper mesosphere. The corresponding meridional velocities (Fig. 9.8) indicate that the upper branch of the global cell in the mesosphere extends sufficiently low to include the 0.3 mbar level during winter and spring, but during summer and autumn the 0.03 mbar and 0.3 mbar levels exhibit oppositely directed meridional motion. In further studying the momentum budget, Crane *et al.* concluded that eddy momentum fluxes due to planetary waves are sufficient to balance the momentum budget of the stratosphere. They play only a minor role in transferring momentum in the mesosphere. Some yet unidentified mechanism is playing an important part.

Fig. 9.8 Zonal mean meriodinal velocity, v (ms^{-1}), at 0.03 and 30mb plotted against latitude for March, June, September and December 1976. Positive values denote northward motion.

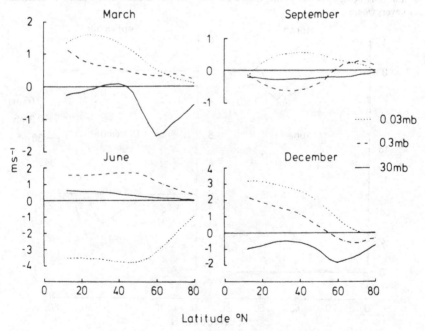

9.4   Wave motions

The most striking features of the time and space variation of the temperature field are the so-called 'sudden warmings' that occur in the winter-time stratosphere and mesosphere. They are best seen in the context of the annual variation of the horizontal distribution of temperature, as illustrated by the sequence of maps in Figs. 9.9 to 9.13. The quantity plotted is the brightness temperature for the B12 channel of the Nimbus 5 SCR (§ 6.8), corresponding to a mean temperature from about 35–50 km. We start with a typical summer situation (Fig. 9.9), with the highest temperature over the pole, as a consequence of solar heating for 24 hours per day. The contours are nearly circularly symmetric, indicating a stable easterly airflow around the pole. After the autumn equinox, Fig. 9.10, the polar stratosphere cools, and the airflow consequently reverses to westerly in accord with the thermal wind equation, but the contours are still fairly symmetric, and the flow is zonal. As the temperature gradient increases, waves develop in the airflow, with consequent variations in the temperature

Fig. 9.9 Brightness temperature from Nimbus 5 SCR channel B12 for day 174, 1973.

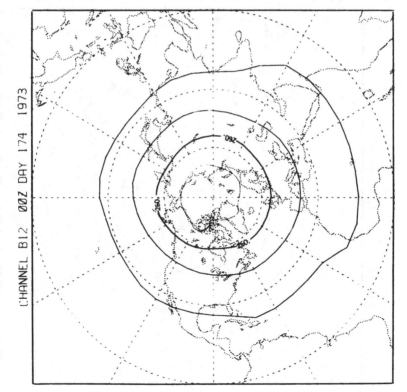

distribution, Fig. 9.11. These waves sometimes become so intense that the orderly flow is completely disrupted (Fig. 9.12), and the temperature disturbance may raise the zonal mean temperature to summer values. After such a sudden warming, the cold pole may re-establish itself, if there is enough time left before the spring equinox (Fig. 9.13). After this time, the polar region is heated by solar radiation, and an orderly easterly flow is established once more.

    The behaviour of the winter stratosphere is only one aspect of wave motions in the middle atmosphere. There are many kinds of waves, most of which can be seen in the temperature distribution from satellites. The main exception is small-scale gravity waves, which are difficult to detect simply because of their scale. Wave motions have various properties that may be used as a convenient classification. They may be local or 'planetary', i.e. global in scale, although of course all waves can in principle be represented as combinations of normal modes of oscillation of the system. Waves may be forced or free; forced waves transfer energy from a source region to a sink region, while free waves are resonant modes

Fig. 9.10 Brightness temperature from Nimbus 5 SCR channel B12 for day 300, 1973.

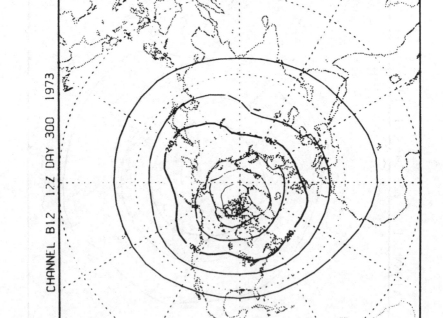

excited by some kind of random forcing. Waves may be travelling or stationary relative to the surface, and may be associated with a zonal wavenumber, indicating the number of oscillations around a latitude circle.

The main classes of wave motion which have been seen in the middle atmosphere from satellites are

(1) low wavenumber (1, 2 or 3) stationary or nearly stationary planetary waves in the winter hemisphere, tilting westward with height, and hence propagating energy upwards. They are probably forced by surface features or slowly moving long waves in the tropospheric weather systems.

(2) Kelvin waves in equatorial regions. These transport energy and momentum upwards from the troposphere, and are probably associated with the semiannual and quasi biennial oscillation in equatorial wind (Hirota 1979).

Fig. 9.11 Brightness temperature from Nimbus 5 SCR channel B12 for day 354, 1973.

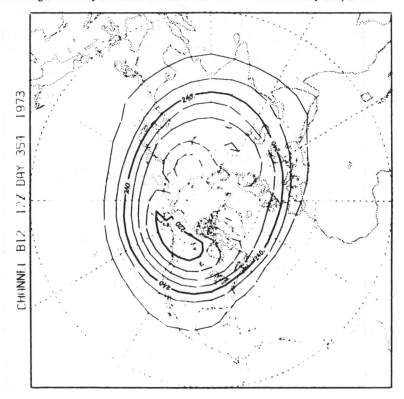

(3) Tides. These are forced by solar heating and to a lesser extent by solar and lunar gravity. They travel westward with a period of one day. They have not been studied from satellites to any great extent because most remote sounders to date have been placed in sun-synchronous orbits.

(4) Free modes. These are usually low wavenumbers (1 to 4), westward travelling planetary waves, with no phase tilt with height or latitude. They do not transport significant amounts of energy or momentum.

9.5   The winter stratosphere

As has been seen, the winter stratosphere is dominated by large amplitude planetary waves, the details of which are still only qualitatively understood. Measurements from satellites are invaluable in providing diagnostic information, such as energy, heat and momentum transports.

These waves originate in the troposphere as a result of the effects of the distribution of continents and ocean, topography, varying cloudiness or varying

Fig. 9.12 Brightness temperature from Nimbus 5 SCR channel B12 for day 57, 1974.

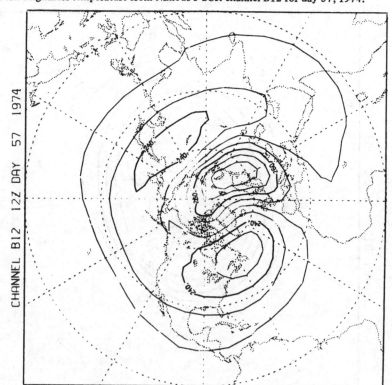

sea surface temperature. The upward propagation of these waves in the atmosphere turns out to be a complex phenomenon dependent on a number of atmospheric parameters.

Charney & Drazin (1961) provided the first theory of this upward propagation. Important extensions to the theory have been provided by Dickinson (1969), Matsuno (1970) and Simmons (1974). The main conclusions of the theory, which are also supported by observations, are:

(1) The structure of the zonal wind, including its variation with height and latitude, is an important parameter in the determination of conditions under which upward propagation is possible.

(2) Upward propagation of planetary waves only occurs when the wave velocity is easterly relative to the zonal flow. This means that upward propagation of waves which are quasi-stationary with respect to the surface does not occur in easterly zonal flow, i.e. such waves cannot propagate under the conditions of the summer circulation.

Fig. 9.13 Brightness temperature from Nimbus 5 SCR channel B12 for day 55, 1973.

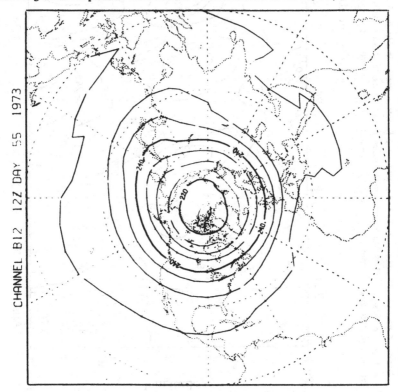

176

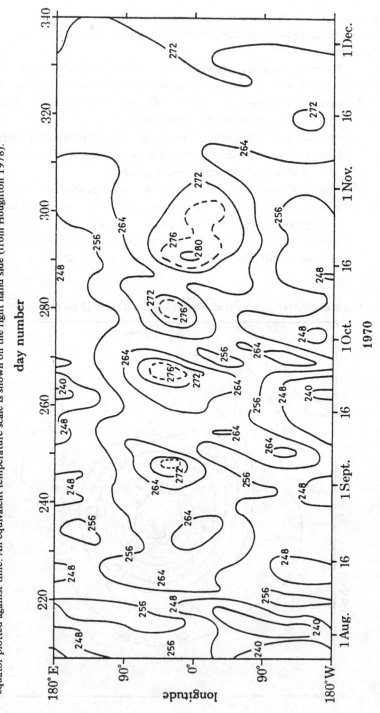

Fig. 9.14 Radiance measurements for the Southern Hemisphere from channel *A* of the Nimbus 4 SCR for the period August to November 1970. Channel *A* peaks at 45 km and possesses a weighting function shown in Fig. 6.13. (*a*) Longitude-time cross section around the 64°S latitude circle. Radiances plotted as equivalent temperatures. (*b*) (next page) Mean radiance averaged around latitude circles at 80°S, 50°S and the equator plotted against time. An equivalent temperature scale is shown on the right hand side (from Houghton 1978).

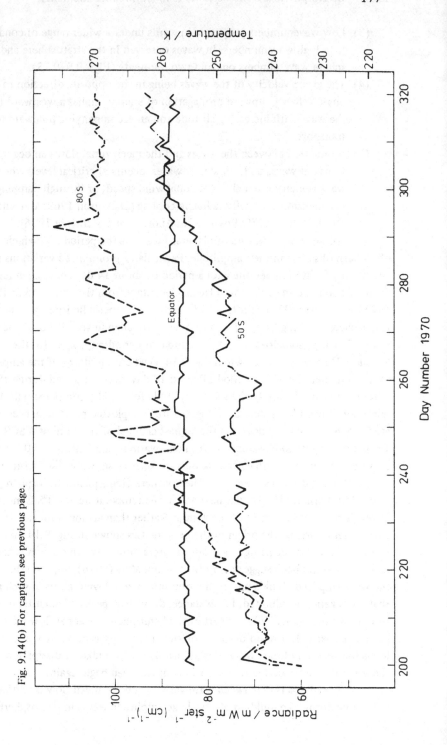

Fig. 9.14(b) Radiance measurements for the Southern Hemisphere from channel A of the Nimbus 4 SCR for the period August to November 1970. Channel A peaks at 45 km and possesses a weighting function shown in Fig. 6.13. (a) Longitude-time cross section around the 64°S latitude circle. Radiances plotted as equivalent temperatures. (b) (next page) Mean radiance averaged around latitude circles at 80°S, 50°S and the equator plotted against time. An equivalent temperature scale is shown on the right hand side (from Houghton 1978).

Fig. 9.14(b) For caption see previous page.

(3) Low wavenumbers propagate upwards under a wider range of conditions than high wavenumbers. In waves observed in the stratosphere and mesosphere wavenumbers one and two dominate (Figs. 9.9-9.13).

(4) The group velocity of the waves being in the opposite direction to their phase velocity, upward propagation of energy implies a westward tilt to the wave with increasing altitude and an accompanying poleward heat transport.

(5) Interaction between the waves and the mean zonal flow can occur for transient waves and, for steady waves, occurs at critical levels where the wave velocity is equal to the zonal wind speed, or through damping mechanisms, especially radiative cooling (Dickinson 1969; Matsuno 1970; Holton 1975; Boyd 1976; Andrews and McIntyre 1976).

An interesting feature of planetary waves is the periodicity which has often been observed in their amplitude. Particularly pronounced variations in the amplitude of wavenumber one with a period of about 14 days have been reported from an examination of Southern Hemisphere data from the Nimbus 5 SCR for 1973 by Barnett (1977). Hirota (1971) suggests that periodic forcing from the troposphere resulting from variations in mean zonal wind velocity could be the cause of such periodicities. Another example of periodicity, again for the Southern Hemisphere, is shown in Fig. 9.14(*a*) where variations of the amplitude of wavenumber one with a period of about 14 days were observed during 1970. After each peak of activity the wave amplitude fell rapidly compared with the relatively slow build-up period. In Fig. 9.14(*b*) are plotted mean radiances averaged around latitude circles from the highest channel of the Nimbus 4 SCR for the period August to November 1970. Polar warming amounting to 10 K or more is evident, associated with each peak of wave activity shown in Fig. 9.14(*a*). Notice also the particular feature of the Southern Hemisphere stratosphere, first noticed by Barnett (1974), that near 45 km the temperature at 50°S is much lower than at 80°S for most of the winter. Rather than arising from variations in tropospheric forcing, the origin of the periodicities shown in Fig. 9.14 may be a relaxation process occurring as a result of interaction between the waves and the zonal flow as has been suggested by Holton and Mass (1976). Suppose that, as the wave amplitude builds up, the distribution of zonal wind alters in such a way that propagation is inhibited. For instance, during the peaks of polar temperature shown in Fig. 9.14(*b*) over part of the hemisphere the zonal flow may just become easterly. If upward propagation of wave energy is prevented, at the upper levels the wave amplitude will die away mainly due to radiative dissipation until the zonal flow is so altered that the whole process can begin again.

The interaction between wave activity and the mean flow is particularly noticeable during the build-up of very large amplitude waves in stratospheric warmings. Matsuno (1971) considered theoretically the interaction which can occur at a critical level where the wave velocity is equal to that of the zonal wind, and showed that if the disturbance is sufficiently persistent and intense, the westerly 'jet' in the stratosphere can be replaced by easterlies. Fig. 9.15 from Matsuno's paper shows the relations between various processes affecting temperature change during a stratospheric warming in the Northern Hemisphere. Away from the critical level the substantial eddy flux of heat towards the pole is approximately balanced by a induced meridional circulation – a result first obtained for planetary waves by Charney and Drazin (1961). Near the critical level, however, the compensation is incomplete and there is a significant net heating at high latitudes balanced by a cooling at low latitudes. Also the Coriolis torque on the southward branch of the meridional circulation decelerates the westerly zonal flow. Matsuno did not consider the effects of eddy momentum transport, but it is easy to see how its divergence may directly affect the local zonal flow (see, for example, Dickinson 1975; Holton 1976). A very close connection exists, therefore, between the eddy transports of heat and momentum and the transport through the mean meridional circulation.

9.6   Free modes

The times and positions of measurements from a vertical sounder are arranged in a way that makes them very convenient for separating out the various modes of oscillation of the middle atmosphere. At any given latitude, a measure-

Fig. 9.15 Illustration from Matsuno (1971) of the meridional circulation induced by the waves showing eddy heat transport $\overline{T'V'}$ components V and W of the mean circulation, and the resultant mean temperature change $\partial T/\partial t$ all near the critical level Z_c.

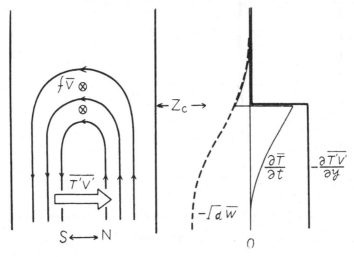

ment is made every orbit (about every 100 mins.) on the ascending side, say, at equal intervals in longitude of about 25°. Thus we obtain a time series of radiances from an observatory which moves westward at 15° longitude per hour, making measurements at about every 25° of longitude. A similar interlaced time series is obtained from the descending part of the orbit. Chapman *et al.* (1974) pointed out that a power spectrum of this time series is a very useful diagnostic tool. Standing waves of wavenumber *n* will appear in the spectrum at *n* cycles per day, due to the Doppler shift between the observatory and the atmosphere, whilst travelling waves of wavenumber *n* will appear at higher or lower frequencies according to whether they are travelling eastwards or westwards.

Chapman *et al.* used this technique to study the eastward travelling planetary waves in the winter. More recently Rodgers and Prata have found that the method is sensitive enough to separate individual travelling waves with amplitudes of a fraction of a degree from the very large amplitudes (tens of degrees) that occur in winter. The waves for which this technique is very useful are the free modes of oscillation, or resonances, which are probably excited by random fluctuations from the tropospheric weather systems. On theoretical grounds these waves are expected to be global in extent, everywhere in phase or antiphase both horizontally and vertically, and to transport no energy or momen-

Fig. 9.16 Power spectrum for May 1973 at 44°N. The main peak is travelling wave with a period of 6.2±0.5 days. The subsidiary peak has a period of 9.2± 1 days. There is no significant standing wavenumber 1. The units of spectral power are K^2 (cycles per day)$^{-1}$.

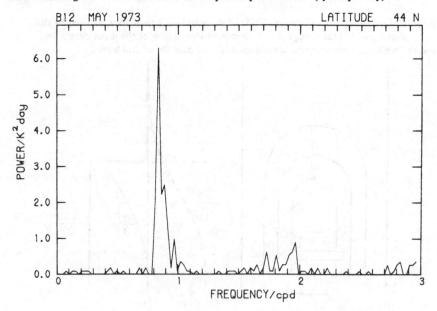

tum. In practice they must carry small amounts of energy and momentum because there are excitation and dissipation mechanisms.

The 'five-day wave' is the wavenumber-one gravest westward-travelling Rossby mode. It has been detected in surface pressure observations at an amplitude of about 0.5 mbar (Madden and Julian 1973). It shows very clearly in spectra of the time series of radiances, with an amplitude of about 0.5 K at 50 km (Fig. 9.16). The amplitudes found agree well with the numerical model of the wave of Geisler and Dickinson (1976), being typically 0.5 K at around 50°N and 50°S, and somewhat less in equatorial and polar regions (Fig. 9.17). The variation with time is quite large, from about 0.2 K to 0.8 K at 50°, and has a time scale of about one month. This is to be expected if the wave is driven by tropospheric weather disturbances.

Fig. 9.17 Amplitude of the five day wave at 1mb as a function of latitude for November 1973.

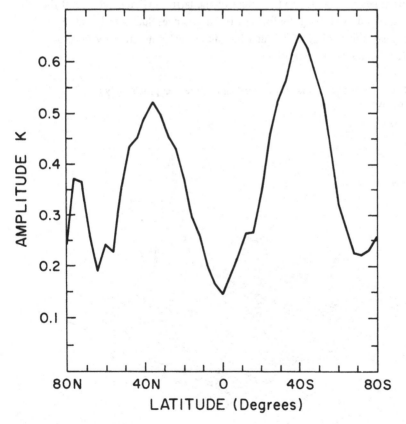

Two other free waves that occur very clearly in the spectra are shown in
Fig. 9.18, using PMR channel 2100 brightness temperatures during January 1976
The peak centred at about 2.52 cpd in the southern hemisphere is interpreted as
a westward travelling wave 3 of local period 2.1 days. The peak at 3.41 cpd is a
westward travelling wave 4, with a period of about 1.7 days. In order to obtain
wavenumber and direction of propagation, the two time series for the ascending
and descending parts of the orbit are filtered at the frequency of the wave and
compared. If the two series have zero relative phase difference, then the oscillations must be of even wavenumber, and if the phase difference is about 180°
then they must be of odd wavenumber.

Variations due to these waves have been seen in meteor winds (Muller
and Nelson 1978), but the global structure can only be seen from satellites. The
amplitude structure, both in meteor winds and radiances, is strongly asym-
metrical with the largest amplitudes at tropical latitudes of the summer hemisphere. This asymmetry may be seen in Fig. 9.18 where a portion of the spectrum
has been plotted against latitude at a time when the amplitude is largest. This
behaviour appears to be due to the effect of the latitudinal dependence of the
zonal mean wind modifying the form of the resonance. Such amplification in
summer solstice winds has been found in a numerical simulation by Prata and
Rogers (Personal Communication).

**Fig. 9.18 Power spectra for January 1976 of temperature at approximately 62km, as a
function of latitude.**

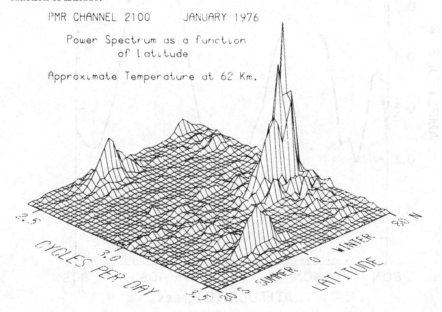

10

MINOR CONSTITUENT ABUNDANCE MEASUREMENTS

10.1 Introduction

The most important gaseous minor constituents in the earth's atmosphere are water vapour, ozone and carbon dioxide. They absorb significant amounts of both solar radiation and terrestrial radiation, ozone notably in the ultraviolet, water vapour and carbon dioxide in the infrared, thus considerably modifying the radiation field and temperature structure within the atmosphere. Because of these strong absorption and emission bands, remote sensing experiments to measure their distribution are possible. Water vapour and ozone possess distributions which are very variable in space and time; because of their significant interaction with atmospheric processes it is required to know their detailed distribution.

Water vapour is especially important in the troposphere because of its role in cloud formation and precipitation and in transporting significant amounts of energy in the form of latent heat. In the stratosphere, in addition to its contribution to the radiation budget, water vapour plays a part in photochemical processes and influences, for instance, the ozone distribution. At higher levels (>80 km) water vapour is dissociated by ultraviolet solar radiation and is the major source of atomic hydrogen.

Ozone is present in the stratosphere with a peak concentration around 25 km, where it is made by photochemical processes. Solar ultraviolet radiation is strongly absorbed by the upper part of the ozone layer resulting in the high temperature region near 50 km altitude. Radiative transfer through the infrared band of ozone is important in the lower stratosphere.

We have already seen (§ 5.1) that carbon dioxide is substantially uniformly mixed up to about 100 km which makes temperature-sounding experiments based on CO_2 emission bands possible. Carbon dioxide plays a dominant role in the energy budget in the mesosphere; the very cold region near the mesosphere results from emission of radiation by CO_2 in its 15 μm band. At these levels thermodynamic equilibrium for this band begins to break down (Houghton 1969); a feature which, as well as being important in the energy budget, affects the interpretation of remote-sounding experiments based on this band.

In recent years a lot of emphasis has been given to catalytic cycles which can significantly change the equilibrium distribution of ozone. Prominent among those have been cycles involving the oxides of nitrogen and chlorine compounds which arise from natural sources or which may be produced as a result of man's activities. Further details of these and other reactions have been given, for instance, by Thrush (1979). In addition to H_2O, O_3 and CO_2 which have already been mentioned, the minor constituents which have so far been measured from satellite instruments are NO, NO_2, N_2O, CO, CH_4, HNO_3. Fig. 10.1

Fig. 10.1 Distribution of some chemically active constituents in the stratosphere, meso-sphere and lower thermosphere (from 'Upper Atmosphere Research Satellite Program', JPL Publication 78-54, 1978).

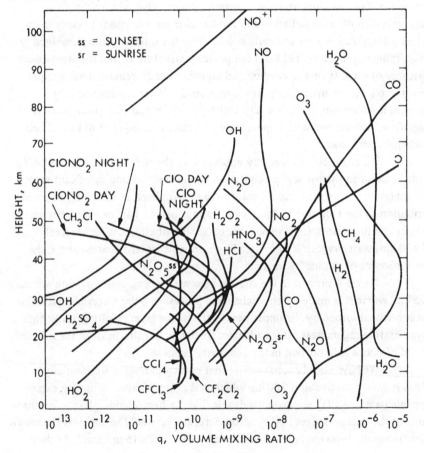

gives an indication of the present state of knowledge regarding the distribution of these constituents with altitude. Measurement of many other constituents is an aim of a number of instruments currently being built for future space missions. The most important of these constituents at the higher levels is atomic oxygen, which contributes to the energy budget of the lower thermosphere for three reasons, namely (1) because it represents a supply of stored chemical energy, (2) because of transfer in its absorption lines in the far infrared and (3) because of its influence on the relaxation of the vibration of carbon dioxide (Crutzen 1970).

Other minor constituents of importance are pollutants present mainly in the lower atmosphere; probably the most important of these is SO_2. Considerable interest is being generated at the present time in the ways in which their distribution can be measured from space. Clouds and aerosols which may also be considered as minor constituents are discussed in Ch. 11.

Remote sounding of minor constituents may be carried out either by observing their absorption of solar radiation transmitted or scattered by the atmosphere or by making measurements on their emission bands in the infrared. The first method has only so far been used for ozone (see § 10.3), the second has mainly been applied to ozone and water vapour. For infrared emission measurements the temperature profile needs to be known, for example by measurements in the $15\,\mu m$ band of CO_2 where the effect of absorption by O_3 or H_2O is relatively small. Then, in principle, additional radiance measurements in regions where the trace constituents are optically active will permit the recovery of vertical profiles of absorber amount. The theory behind this is complicated by the fact that there are few spectral regions where the opacity can reasonably be taken as due to one absorber alone. Absorption by H_2O is not negligible in the region of both the $15\,\mu m$ CO_2 band which is used for temperature sounding and the $9.6\,\mu m$ O_3 band which is the most convenient for ozone sounding. Water vapour lines also contaminate the atmospheric windows around $10\,\mu m$ which are used to sound the surface temperature (§3.2.3).

10.2 Water vapour

The MRIR on early TIROS satellites (§ 4.3) possessed a channel in the $6.3\,\mu m$ water vapour band in order to estimate atmospheric humidity. Since the saturation vapour density and temperature are uniquely related, in the absence of cloud, for a constant humidity, the optical depth of water vapour will be a function of temperature only. This means that under conditions of constant humidity, a roughly constant radiating temperature will be observed in the water vapour band. Variations of radiating temperature, therefore, arise from variations in the humidity over an altitude range broadly representing the upper troposphere. Comparison with the window region (8–12 μm) needs to be made to eliminate the effect of cloud – a procedure which was strongly criticized by Fritz and Rao (1967) who pointed out the strong variation of cloud properties with wavelength in this part of the infrared.

　　More refined measurements of water vapour distribution have been made from the IRIS instruments on Nimbus 3 and 4 (Conrath 1969; Prabhakara *et al.* 1979) and from the SIRS instrument on Nimbus 4 which included channels in the rotation band of water vapour between 20 μm and 40 μm. Fig. 10.2 shows

Fig. 10.2 Weighting functions for six channels of the SIRS on Nimbus 4 observing in the rotation water vapour band (from Smith 1970).

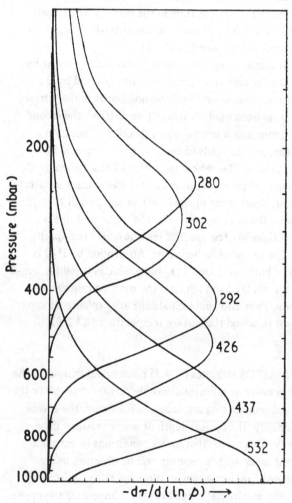

a set of weighting functions for the six channels of the SIRS instrument computed for a standard atmosphere at a constant relative humidity of 40%. A retrieved water vapour profile from this instrument is shown in Fig. 10.3.

Hillger and Vonder Haar (1977) have demonstrated, using measurements from the VTPR on NOAA satellites (§ 6.6), that by combining satellite and radiosonde observations of tropospheric water-vapour observations can be achieved showing a lot of detail in their horizontal coverage and, therefore, having important applications in mesoscale meteorology. Further studies of the retrieval of moisture fields from the HIRS instrument (para 6.6) on the Nimbus 6 and on the TIROS N satellites have been made by Hillger and Vonder Haar (1981) and Hayden *et al.* (1981).

Information on the distribution of upper tropospheric water vapour is available from Meteosat (Eyre 1981) (Fig. 12.2). The VAS (para 6.6) now mounted on the GOES geostationary satellites also possesses channels for the measurement of water vapour, in particular the 'split window' channels near 11 μm provide information about water vapour in the lowest few kilometres of the troposphere.

Fig. 10.3 Comparison between water-vapour profile retrieved radiance observations from SIRS on Nimbus 4 (solid line) over 13°N 127°E at 1532 GMT on 8 April 1970 with radiosonde observations at 15°N 121°E at 1200 GMT (short dashes) and 1800 GMT (long dashes) on the same day. The circles show the first guess profile (from Smith 1970).

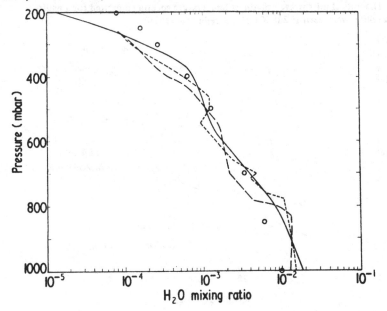

Water-vapour measurements may also be made by observing in the microwave part of the spectrum. Early measurements from the Cosmos 243 satellite (Bosharinov *et al.* 1969; Gurvich and Demin, 1970) determined total water-vapour content over the oceans from observations at 1.35 cm wavelength. Coincident measurements at 0.8 cm were used to eliminate the effect of clouds. Fig. 10.4 shows computed spectra for liquid water and for water vapour.

Staelin *et al.* (1976) have employed measurements from the NEMS on Nimbus 5 (§ 6.9) at frequencies near 22 and 31 GHz to infer the total water-vapour and total liquid-water content of the atmosphere. A water-vapour resonance line occurs at 22 GHz around which liquid water has a broad absorption. By measuring at the two frequencies, therefore, both liquid water and water vapour can be inferred. Fig. 10.5 shows a scan across a region near a tropical cyclone. Notice the warm core of the storm as indicated by the brightness temperature of one of the NEMS temperature sounding channels which monitors the region between about 7 and 15 km altitude. The SCAMS on Nimbus 6 combined spectroscopy in the same channels as NEMS with scanning across the orbital track so that images of water-vapour and liquid-water content could be produced (Staelin *et al.* 1977). In the interpretation of microwave measurements from the two channels we have mentioned, assumptions have to be made regarding the emissivity of the sea surface – a quantity which is dependent on sea state (cf Fig. 3.10): Rosenkranz *et al.* (1978) discuss methods for dealing with this

Fig. 10.4 Theoretical microwave brightness temperature spectra computed for a nadir-viewing satellite over ocean at 288 K (after Staelin *et al.* 1976).

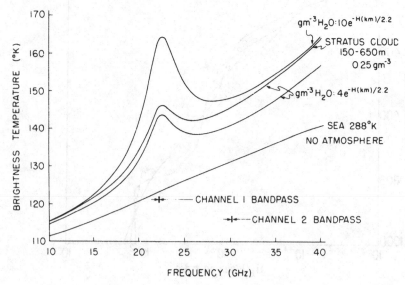

problem. Maps of the total water vapour content over the oceans have been produced from the Nimbus 6 SCAMS (Grody *et al.* 1980) and from a similar instrument, the Scanning Multichannel Microwave Spectrometer (SMMR) on Nimbus 7
(Prabhakara *et al.* 1982 and Fig. 10.6). Very good agreement between satellite
observations and radiosonde measurements is achieved. (Fig. 10.7). For microwave observations of water vapour profiles, measurements can be made near
183 GHz water vapour line (Rosenkranz *et al.* 1982, Fig. 6.27 and Fig. 10.8).
The AMSU (para 6.9) which is being planned for future operational meteorological observations will contain a number of high frequency channels for this
purpose.

In the higher atmosphere the water-vapour distribution may be observed
by limb-sounding radiometers. The great advantage of limb-sounding geometry,
as was mentioned in § 5.1, is the long path of emitting atmosphere behind which
is the cold background of space. When viewing the limb, the optical path above
the tangent point is about 70 times the vertical path above the same level.

Both LRIR and LIMS possess channels in the 6.3 μm water-vapour
band and can probe stratospheric water vapour up to about 50 km altitude. The
problem of measuring above that level is that the emission even of the long
path as measured by a filter radiometer in a water-vapour band becomes very
low – less than 1% of the emission from a black-body at instrument temperature – and difficult to distinguish from stray radiation arising from within the

Fig. 10.5 Retrievals of total atmospheric water vapour and liquid water from the Nimbus
5 microwave spectrometer (NEMS) along a path crossing a tropical cyclone in the southern
Indian Ocean on 22 February 1973 (after Staelin *et al.* 1976).

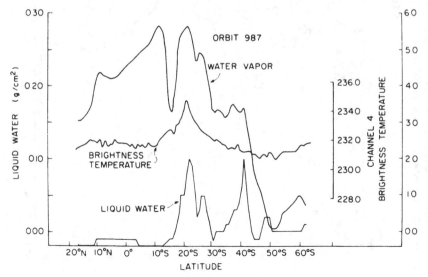

190

Fig. 10.6 Distribution of precipitable water in the atmosphere over the global oceans derived from Nimbus 7 SMMR data for the period 25 October - 25 November 1978 (after Prabhakara *et al.* 1982).

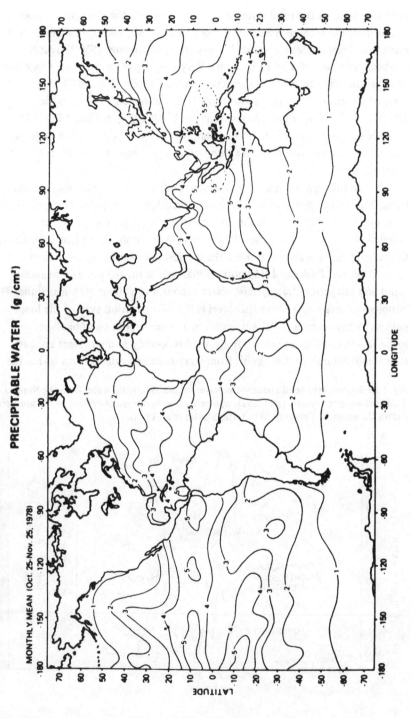

radiometer itself. One way of overcoming this problem is to select radiation from the emission lines themselves by using the technique of gas correlation spectroscopy described in § 6.7. This has been done in an instrument known as the Stratospheric And Mesospheric Sounder (SAMS) flown on the Nimbus 7 satellite in 1978.

Table 10.1 lists the channels included in SAMS and Fig. 10.9 illustrates the optical system. Mirror M1 scans over the atmospheric limb; mirrors M2 and M3 constitute a telescope providing an image of the limb on the faceted mirror M5. Three parts of this field are separated by M5 and directed through various pressure modulator cells onto two cooled detectors labelled A234 and B1 and to uncooled detectors labelled A1, B2 and C2/3. A 'shallow' mechanical chopper (i.e. a chopper that only chops part of the aperture) is included to enable signals (called wide-band signals) such as would be appropriate to simple filter channels to be obtained from each detector. Six pressure modulators are included in SAMS. In all of them the pressure in the cells may be altered in flight so that measurements may be made over a wide range of altitudes (*cf.* Fig. 10.10).

Fig. 10.7 Comparison of ship radiosonde measured precipitable water with that estimated from Nimbus 7 SMMR data (after Prabhakara *et al.* 1982).

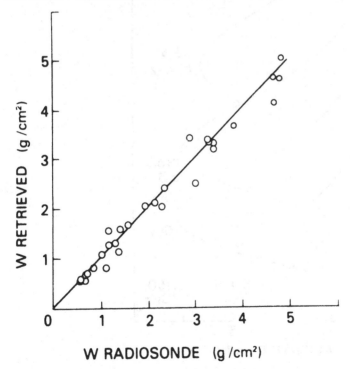

Because SAMS is a limb-viewing instrument, it is necessary that accurate information is available regarding the pressure of the atmosphere at the levels being viewed at any given time. Consider information such as that contained in Fig. 10.10. When, for instance, the effective emissivity of an atmospheric path (see p. 73) is about 0.5, a change in signal corresponding to a change in emissivity

Fig. 10.8 Weighting functions for the high frequency channels of the Advanced Microwave Sounding Unit (from Rosenkranz *et al.* 1982).

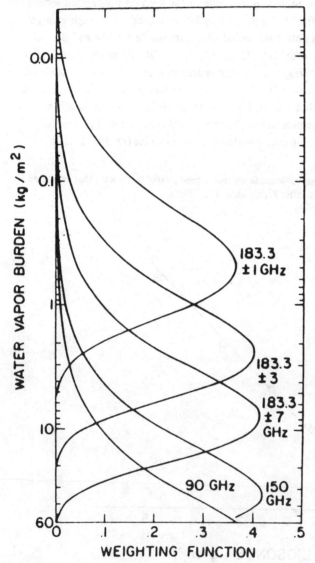

of 0.01 (equivalent to a change in mixing-ratio of 6%) could also result from a change in the level being viewed of 0.1 km, which is equivalent to a change in pressure at the tangent point of ~2% or 0.003° in viewing direction. The attitude of the spacecraft is controlled only to ~0.5° in all three axes; it is, therefore, necessary from the measurements made by SAMS itself to determine the appropriate information about the pressure at the part of the atmosphere being observed.

For the LRIR and LIMS instruments, attitude information is obtained as a result of the retrieval process (Gille and House 1971). Essentially the same method is employed in SAMS by comparing the signals from a wide-band and a pressure modulator channel in the 15 μm CO_2 band. Under the approximation of equation (5.13) the two signals will be $\epsilon_w B(\bar{T}_w)$ and $\epsilon_p B(\bar{T}_p)$ where the sub-

Fig. 10.9 Optical schematic of the stratospheric and mesospheric sounder (SAMS) on Nimbus 7 (after Drummond *et al.* 1980).

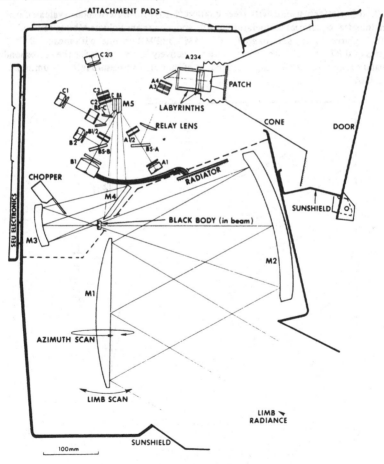

scripts w and p respectively denote the wide-band and pressure modulator channels. Now it is possible to choose the mean cell pressure in the pressure modulator channel such that over a certain range of altitudes the relative contributions
of different segments of the atmospheric path being observed are very similar for
the two channels even though the effective emissivity of the atmosphere for the
two channels is very different (Fig. 10.11). This means that $\bar{T}_w$ and $\bar{T}_p$ will be
almost equal. We then find that the ratio ϵ_p/ϵ_w is almost independent of temperature but strongly dependent on the pressure at the level of observation.
Fig. 10.12 illustrates that, for a particular cell pressure, atmospheric pressure
over the range between 0.8 mbar and 20 mbar may be derived from the ratio of
signals in the two channels. A first order correction for the atmospheric temperature profile can be included enabling pressure measurements of adequate accuracy
($\sim$2%) to be achieved.

Fig. 10.10 The effective emissivity (see equation (5.13)) for water-vapour evaluated using
a climatological equatorial temperature profile and a constant volume mixing ratio of
5×10^{-6}. Curves (*a*) (*b*) and (*c*) are for the SAMS B2 PMR channel with mean cell
pressures of 0.87 mb, 4.48 mb and 15.3 mb respectively. Curve (*w*) is for the corresponding
wideband channel and is virtually independent of modulator pressure (after Drummond
et al. 1980).

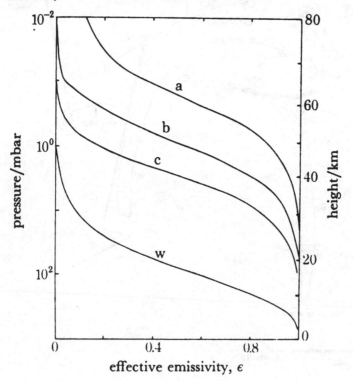

Table 10.1. *SAMS radiometric channels (Drummond et al. 1980)*

Field of View	Channel	Constituent (Gas in modulator)	Mean cell pressures (mbar)	Spectral band (μm)	Derived quantities & altitude range
A	A1	CO_2	17, 2.4, 0.69, 0.25	15	Kinetic temperature 15–80 km; Attitude
	A2			4.3	
	A3	CO	14.8, 4.5	} 4–5	Distribution 15–60 km
	A4	NO	45, 20		
B	B1	H_2O	16, 4.5, 0.8, 0.5	2.7	Distribution 80–100 km
	B2			25–100	Distribution 15–100 km
C	C1	CO_2	36, 11.2, 3.25, 0.87	15	Kinetic temperature 15–80 km; Attitude
	C2	N_2O	24.4, 7.15	} 7.7	Distribution 15–60 km
	C3	CH_4	47.8, 22.5		

Most of the SAMS observations are of thermal emission from the various constituents listed in Table 10.1. From the near infra-red bands resonance fluorescence of solar radiation can also be observed (Fig. 10.13). In particular, resonance fluorescence from the 2.7 μm band of H_2O has been interpreted in terms of the water vapour distribution in the mesosphere (Fig. 10.14). Although the signal: noise for this measurement is low, the results show a volume mixing ratio of about 10^{-6} at 70 km altitude falling to much lower values above 80 km.

10.3   Ozone

Ozone possesses a strong absorption band in the ultraviolet below about 0.3 μm. Measurements of the total ozone content of the atmosphere and of its distribution have been made from the ground for many years using the technique pioneered by Dobson which involves the measurement of the attenuation of sun-

Fig. 10.11 A typical pair of CO_2 15μm weighting functions for the SAMS attitude-determining channels before the finite field of view is accounted for. The tangent height of the line of sight is 50 km.

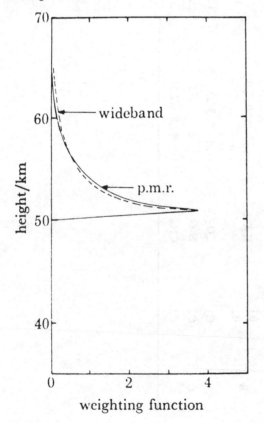

weighting function

Fig. 10.12 Ratio of effective emissivities $\epsilon_\rho/\epsilon_\omega$ for 15μm CO_2 channels on SAMS. The shaded area denotes the range of curves derived using different atmospheric temperature profiles (after Drummond *et al.* 1980).

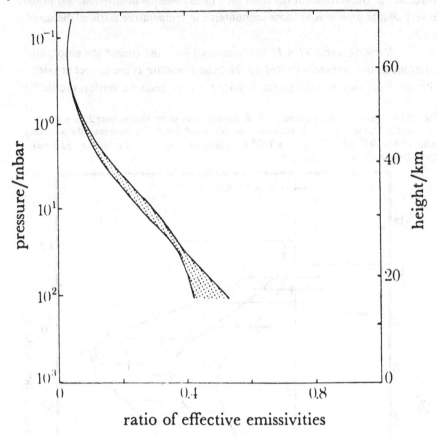

ratio of effective emissivities

Fig. 10.13 Illustrating resonance fluorescence observation.

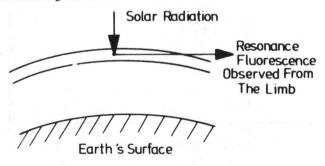

light at different ultraviolet wavelengths. Dobson's work on ozone was one of
the first remote-sounding experiments on the atmosphere, and the theory of the
deduction of the vertical profile from such measurements involved Mateer (1965)
in very similar problems to those encountered in temperature retrieval discussed
in Ch. 7.

   Venkateswaran *et al.* (1961) measured from the ground the absorption
by ozone in the sunlight reflected by the Echo 1 satellite as the sun set at satellite
altitude; from such measurements they were able to deduce a vertical profile for

Fig. 10.14 Measurements of resonance fluorescence from water vapour near 2.7μm from
the Nimbus 7 SAMS. The three solid lines are calculated signals for constant volume mixing
ratios of 5 x 10^{-7} (*a*), 10^{-6} (*b*), 2 x 10^{-6} (*c*). Horizontal and vertical error bars are shown
(from Drummond and Mutlow 1981).

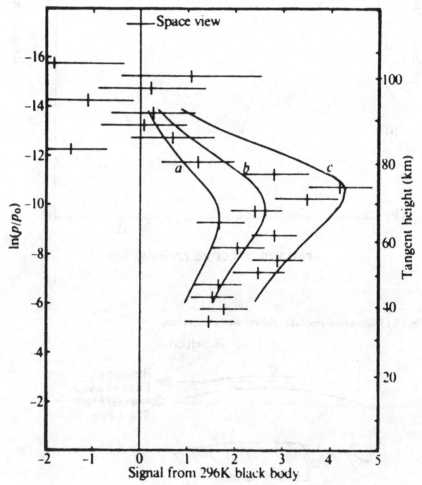

ozone concentration. Similar experiments have been carried out from satellites by Miller and Stewart (1968), Rawcliffe *et al.* (1963) and Guenther *et al.* (1977). Obtaining a vertical profile of ozone from the measurements was complicated by the finite size of the sun. Also, although a substantial number of sunrises and sunsets occur at the satellite during the course of a day, nothing like global coverage is obtained and the observations are all made under identical conditions of illumination of the observed layer.

Similar occultation measurements using stellar sources rather than the sun have been made by Hays & Roble (1973) and Riegler *et al.* (1976). Some results from occultation measurements are presented in Fig. 10.15 where the existence of significant disagreement between different measurements, probably due to instrumental reasons or to faults in interpretation, is demonstrated. Table 10.2 provides a chronology of these along with other satellite techniques for ozone measurement.

The backscatter ultraviolet technique measures the spectrum of reflected sunlight in the ultraviolet and so enables the ozone distribution to be inferred. To illustrate the method, following Twomey (1961) we present a simplified theory on the following assumptions: (1) the sun is overhead, (2) the direction of view is vertically downwards and (3) the atmosphere is infinitely deep so that the ground can be neglected.

At a given wavelength λ the flux of solar radiation at the top of the atmosphere is $S_\lambda(0)$ and at a level where the pressure is p will be

$$S_\lambda(p) = S_\lambda(0) \exp \left(-k_\lambda \int_0^p n(p') \, dp' \right) \tag{10.1}$$

where k_λ is the absorption coefficient per molecule of ozone which is present in concentration $n(p')$ molecules per unit volume at the level of pressure p'. In writing equation (10.1) the further assumption has been made that attenuation of the solar radiation by absorption due to ozone is very much greater than that due to scattering.

From a layer of thickness dp at level p the scattered radiation in the vertically upwards direction will be proportional to $S_\lambda(p)$ and to dp, say it is $aS_\lambda(p) \, dp$. On traversing the return path to the satellite the radiation will be further attenuated by the same amount as in equation (10.1) so that a total scattered radiance L_λ leaving the atmosphere will be

$$L_\lambda = S_\lambda(0) \, a \int_0^p \exp \left(-2k_\lambda \int_0^p n \, dp \right) \, dp. \tag{10.2}$$

200

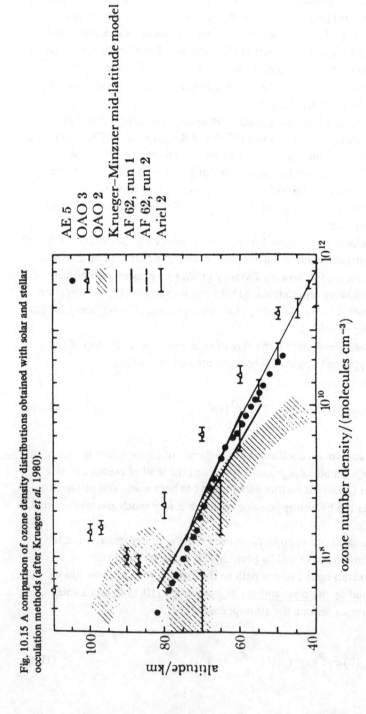

Fig. 10.15 A comparison of ozone density distributions obtained with solar and stellar occulation methods (after Krueger et al. 1980).

If the equation is written in terms of the total number of ozone molecules above a given level, i.e.

$$x = \int_0^P n \, dp \tag{10.3}$$

and the independent variable taken to be $\ln p$, we obtain

$$L_\lambda = S_\lambda(0) \, a \int_0^\infty p \, e^{-2k_\lambda x} \, d(\ln p) \tag{10.4}$$

It is instructive to investigate at what levels the radiation originates. If we suppose uniform mixing ratio, that is, $n = $ constant, then (10.4) can be written

$$L_\lambda = S_\lambda(0) \, a \int_0^\infty p \, \exp\left(-2k_\lambda np\right) d(\ln p). \tag{10.5}$$

The function $p \exp(-2k_\lambda np)$ is very similar to the weighting function of equation (5.5) which is plotted in Fig. 5.2 but with p instead of p^2. It has a peak at $p_{max} = (2k_\lambda n)^{-1}$. Therefore, by choosing a set of wavelengths where k_λ is different, the ozone concentration at different levels can be monitored.

The backscatter ultraviolet spectrometer (BUV) on Nimbus 4, although not the first instrument to employ the technique (*cf.* Table 10.2), was the first such instrument dedicated to ozone measurement to fly on a polar-orbiting satellite. It consists (Heath *et al.* 1970) of a double Ebert-Fastie monochromator selecting twelve narrow wavelength bands in the $0.25-0.34\,\mu m$ region to view reflected sunlight or reflected moonlight from the nadir. Reflection from the lower atmosphere outside the ozone band is monitored by an additional filter photometer at $0.38\,\mu m$ so that lower atmosphere effects can be eliminated. Scattered light within the instrument is reduced by using a double monochromator; calibration is effected either by an internal radioactive phosphor source or by introducing in front of the instrument on command lambertian diffusing plates. The detector is a photomultiplier feeding into an electronic system having a dynamic range approaching 10^9 to cope with the variation of input light level over the different wavelengths, allowing also for the difference between moonlight and sunlight. Details of the retrieval method in which allowance has to be made for varying geometry at different orbital positions of the satellite, multiple scattering, variations of surface albedo, and the possible presence of clouds and aerosol, are given by Dave and Mateer (1967), Herman and Yarger (1969) and Dave and Furakawa (1967).

The total ozone for the period 1970-72 as deduced from the BUV is shown in Fig. 10.16. An example of the correlation between ozone concentration

Table 10.2. *Satellite experiments to measure ozone*

Type	Satellite	Wavelengths	Latitude coverage	Comments	References
Occultation					
Solar	Echo 1	590, 529.5 nm	17°N	Dec. 1960	Venkateswaran et al. 1961
	US Air Force	260 nm	33°S–13°S	July 1962	Rawcliffe et al. 1963
	Ariel 2	200–400 nm	50°S–50°N	April, May 1964	Miller and Stewart, 1965
Stellar	AE-5	225.5 nm	5°N	Dec. 1976	Guenther et al. 1977
	OAO-2	250 nm	16°S–43°N	Jan. 1970	Hays and Roble 1973
	OAO-3	258, 343 nm	12°S–3°N	Aug. 1971 July 1975	Riegler et al. 1976
Backscatter UV					
Profile	USAF 1965	284 nm	60°S–60°N	Feb–Mar 1965	Rawcliffe and Elliot 1966
	USSR	225–307 nm	60°S–60°N	April 1965	Iozenas et al. 1969
		250–330 nm	60°S–60°N	June 1966	Iozenas et al. 1969
	1966-111B	175–310 nm	80°S–80°N	1966	Elliott et al. 1967
	OGO-4	110–340 nm	80°S–80°N	Sept. 1967	Anderson et al. 1969
	Nimbus 4 BUV	225.5–305.8 nm	80°S–80°N	April 1970 July 1977	Heath et al. 1973
	AE-5 BUV	255.5–305.8 nm	20°S–20°N	Nov. 1975– April 1977	Frederick et al. 1977a
Total	Nimbus 7 SBUV	255.5–305.8 nm	80°S–80°N	Nov. 1978	Heath et al. 1975
	Nimbus 4 BUV	312.5–339.8 nm	80°S–80°N	April 1970– July 1977	Mateer et al. 1971
	AE-5 BUV	312.5–339.8 nm	20°S–20°N	Nov. 1974– April 1976	
	Nimbus 7 TOMS	312.5–339.8 nm	global	Nov. 1978	Heath et al. 1975

Type	Satellite	Wavelengths	Latitude coverage	Comments	References
Infrared emission					
Profile	Nimbus 6 LRIR	9.6 μm	65°S–90°N	June 1975–Jan. 1976	Gille 1979
	Nimbus 7 LIMS	9.6 μm	65°S–90°N	Oct. 1978	Nimbus Project 1978
Total	Nimbus 3 IRIS	9.6 μm spectral scan	80°S–80°N		Hanel *et al.* 1970
	Nimbus 4 IRIS		80°S–80°N	April 1970–Jan. 1971	Prabhakara *et al.* 1976
	Block 5 MFR (4 flights)		global	March 1977–	Lovill *et al.* 1978
	TIROS-N HIRS	9.71 μm	global	Nov. 1978–	

204

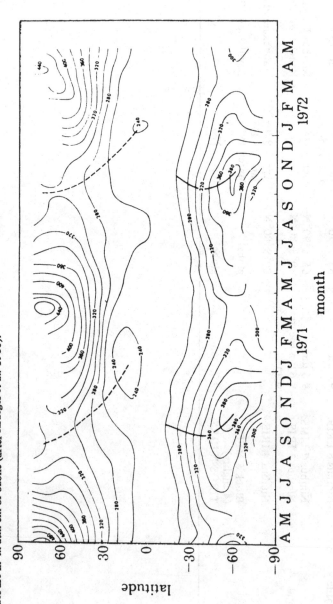

Fig. 10.16 Time latitude cross-section of zonally averaged total ozone obtained from the Nimbus 4 BUV from April 1970 to May 1972. The values are in m atm-cm of ozone (after Krueger *et al.* 1980).

near to 50 km and temperature is shown in Fig. 10.17; the amount by which ozone concentration varies with temperature provides clues regarding the photochemical processes involved (Barnett *et al.* 1975). A particularly important measurement with the BUV was of a sudden decrease in stratospheric ozone coincident with a

Fig. 10.17 Temporal variations of the 2 mbar temperature from SCR and the 1 mbar ozone mass mixing ratio in a latitude band centered at 60 North from one year of Nimbus 4 data (after Krueger *et al.* 1980).

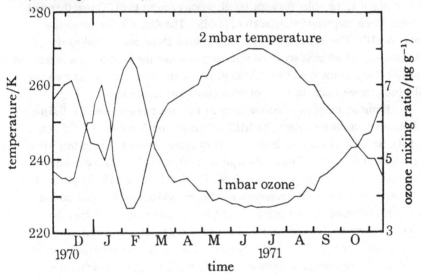

Fig. 10.18 Total ozone above 4 mbar averaged over latitudes 75-80°N for August 1972 as measured by the back-scatter ultraviolet spectrometer on Nimbus 4, after Heath *et al.* (1977). Notice the large drop in ozone content near 4 August when a major solar proton event occurred.

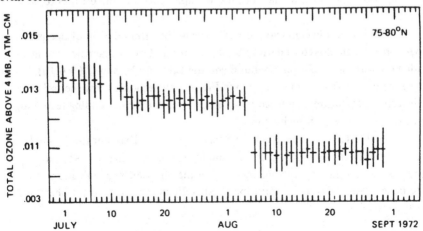

major solar proton event on 4 August 1972 (Fig. 10.18). The connecting mechanism postulated by Heath *et al.* (1977) and confirmed in a model experiment by Fabian *et al.* (1979) is of enhanced nitric oxide concentration due to solar protons incident on the thermosphere in polar regions being transported downwards into the stratosphere where at levels above 4 mbar about 20% of the ozone was destroyed.

The most recent BUV experiment is on Nimbus 7 and is made up of two instruments, the Solar Backscatter ultraviolet sensor (SBUV) together with the total ozone mapping spectrometer (TOMS). The SBUV is similar to the Nimbus 4 BUV. The TOMS employs a single monochromator sampling the backscattered radiation at six fixed wavelengths while the field of view is scanned perpendicularly to the satellite's orbital plane, so enabling continuous mapping of the total ozone since the scans of consecutive orbits overlap.

Remote sounding of ozone using its infrared emission band at 9.6 μm has been made using data from the IRIS instrument on Nimbus 4 (§ 6.5 and Fig. 5.4). Because of the very large number of closely spaced lines in this band, the amount of variation of mean absorption coefficient through the band is rather small. Sekihara and Walshaw (1969), Prabhakara *et al.* (1970) and Shafrin (1970) have shown that the radiances in different regions of the band do not actually contain much information about the O_3 profile, although they do allow quite accurate determinations of the total O_3 amount in a vertical column, itself a useful quantity. A further difficulty is that significant observations can only be made when the stratosphere is at a different temperature from the underlying surface; measurements over high cloud or over Antarctica are liable, therefore, to be very inaccurate. Prabhakara *et al.* (1970) have shown that a statistical, one-parameter approach, which makes use of climatological data on ozone distribution, enables crude but useful approximate ozone profiles to be extracted.

Limb-sounding emission observations in the infrared band of ozone at 9.6 μm may be employed to derive the ozone profile. Ozone channels were included on both the LRIR on Nimbus 6 and the LIMS on Nimbus 7 (§ 6.10). An ozone profile from LIMS is shown in Fig. 10.19, demonstrating the rather high vertical resolution which can be achieved from the limb-sounding technique compared with the BUV technique.

In 1981, NASA launched the Solar Mesosphere Explorer (SME) satellite particularly to study ozone. It is a spinning spacecraft (Barth 1981) carrying ultraviolet and infrared spectrometers to scan the atmosphere's limb and so to determine ozone, NO_2 and water vapour generally in the upper stratosphere and lower mesosphere.

10.4 Other gaseous constituents

The methods described above for sounding for water-vapour and ozone concentration may be applied to other constituents. Many of the minor constituents of importance are present in only a few parts in 10^9. Because of this their spectral signatures are not very clear and are very easily obscured due to the overlap of bands of more abundant constituents.

Because of the long limb path, limb-sounding techniques are the obvious ones to apply to the measurement of minor constituents. LIMS (§ 6.10) has channels for observing NO_2 and HNO_3 and SAMS (§ 10.2) has channels for observing CH_4, N_2O, CO and NO. Some early results from these instruments are presented in Figs. 10.20–10.23.

Nimbus 7 data from LIMS and SAMS are providing information about the distribution in space and time of these constituents. Comparison between model calculations and these observations will lead to an increased understanding of middle atmospheric chemistry and also, because each constituent acts as

Fig. 10.19 Preliminary LIMS ozone retrievals. Solid line, 109°E (day); dashed line, 111°W (night) (after Gille *et al.* 1980).

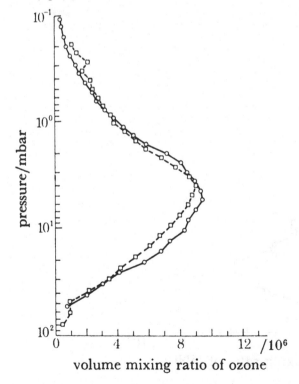

a tracer of atmospheric motion, of dynamical transport processes in the middle atmosphere.

10.5   Future instruments

Already plans are underway for mounting improved models of instruments such as SAMS on future satellite missions, in particular on UARS (Fig. 2.6). Further advances will be achieved by the employment of very high resolution spectroscopy on the one hand and of cooled instrumentation on the other.

An example of a very high resolution instrument is the high-speed interferometer described by Toth and Farmer (1971) which is a Fourier device for observing solar radiation transmitted through the atmospheric limb covering the spectral range 1–5 μm (1000–2000 cm^{-1}) with a resolution of 0.12 cm^{-1}. This instrument has already flown on balloons and aircraft and a version called ATMOS (Atmospheric Trace Molecular Spectroscopy) is planned for the payload of Spacelab 3 in 1984.

Fig. 10.20 Preliminary LIMS nitric acid retrievals. Solid line, 109°E (day); dashed line, 111°W (night) (after Gille *et al.* 1980).

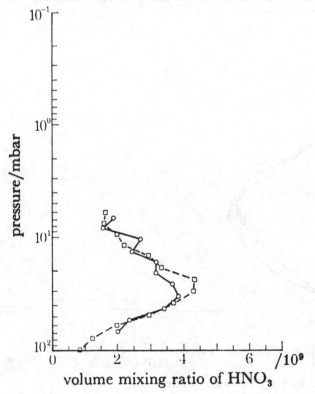

volume mixing ratio of HNO$_3$

Microwave radiometry observing the atmospheric limb may also be employed for minor constituent determinations. Many constituents of interest (e.g. O_3, H_2O, NO, ClO, etc.) possess rotational transitions in the microwave region. To achieve adequate vertical resolution at the atmospheric limb despite the problems of diffraction at these wavelengths, an antenna of at least 1 m diameter is required (Waters *et al.* 1975); there will, however, be little difficulty in mounting such large instruments on the space shuttle or on free-flying satellites in the future.

Fig. 10.21 Preliminary LIMS nitrogen dioxide retrievals. Solid line, 109°E (day); dashed line, 111°W (night). A large diurnal variation is expected (after Gille *et al.* 1980).

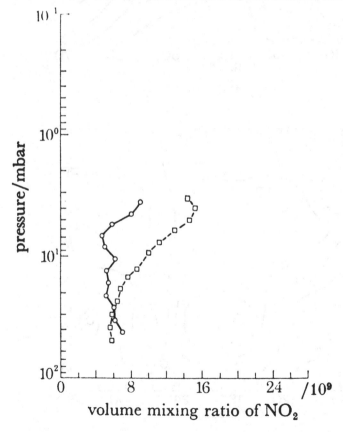

Fig. 10.22 Methane observations from the SAMS Nimbus 7. (*a*) Zonal mean cross-section for 5 February 1981 of volume mixing ratios of CH₄ in ppm. (*b*) Cross section along circle of latitude at 65°N of deviations from zonally averaged mean which were associated with a large amplitude planetary wave.

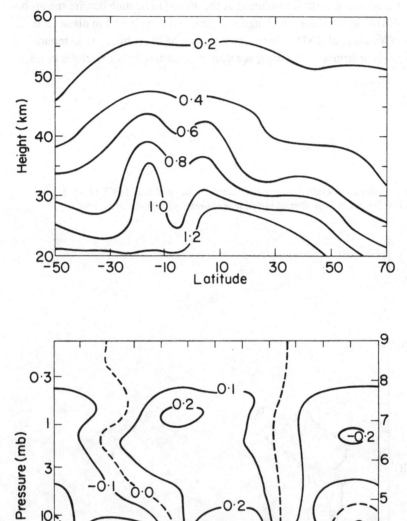

Fig. 10.23 Nitrous oxide observations from the SAMS on Nimbus 7 for 23 February 1979. (*a*) Zonal mean cross section of volume mixing ratio in ppb. (*b*) Cross section along circle of latitude at 65°N of % deviations from zonally averaged mean which were associated with large amplitude planetary wave of wave number 2.

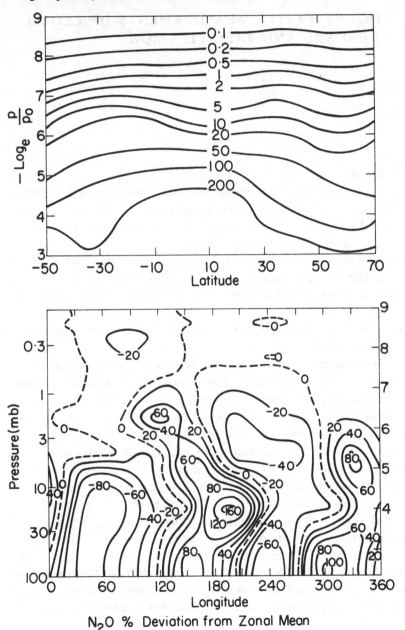

N$_2$O % Deviation from Zonal Mean

11

QUANTITATIVE MEASUREMENTS OF CLOUD, AEROSOL AND PRECIPITATION

11.1   Clouds

Cloud mapping in the visible and infrared has already been described in Ch. 3. Further information about clouds is required for several reasons, namely:

(1) so that the effect of clouds in the interpretation of temperature sounding experiments can be eliminated (Ch. 8);

(2) to assist in the interpretation of radiation budget measurements and the resolution of the cloud/radiation feedback problem (Ch. 4);

(3) to provide information to assist numerical modellers in the parameterization of cloud; and

(4) so that areas of precipitation can be located and estimates of precipitation made.

For many of the purposes we have mentioned, cloud height is an important parameter. Since uniform cloud of sufficient thickness radiates substantially as a black-body, the cloud-top temperature for such a cloud layer may readily be determined from infrared observations. If the atmosphere's temperature structure is known, the height of the cloud-top follows. Many clouds observed from satellites do not, however, follow such a simple specification. Reynolds and Vonder Haar (1977) have made use of simultaneous infrared and visible satellite radiance data in the determination of cloud height. McCleese and Wilson (1976) and Smith and Platt (1978) have employed a number of channels of temperature sounding instruments (the SCR and the ITPR on Nimbus 5 respectively) to determine cloud heights and amounts in addition to the temperature profile. Comparisons made by Smith and Platt (1978) of cloud estimates from the ITPR with ground observations made by a lidar showed good agreement between the two. Another method of cloud-height determination is to observe stereoscopic pairs of images in regions where geostationary satellite images overlap.

Fig. 11.1 The relationships between brightness and cloud thickness (mean curve) (from Park *et al.* 1974).

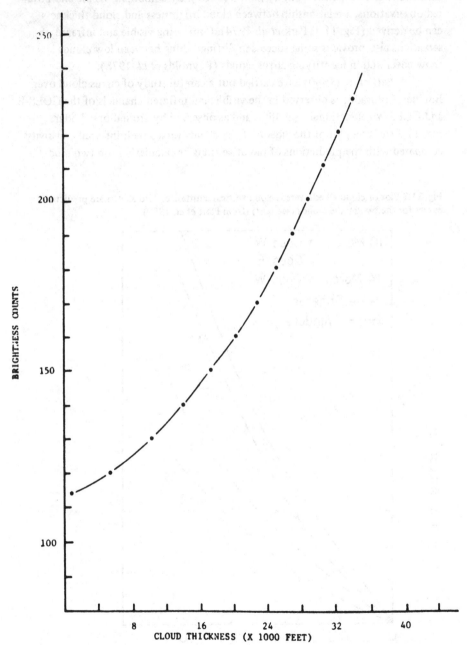

That variations in cloud brightness in the visible are related to cloud thickness was first pointed out by Suomi (1969). By combining visible and infrared observations, a relationship between cloud brightness and cloud thickness can be derived (Fig. 11.1; Park *et al.* 1974). Comparing visible and infrared observations also provides some success in distinguishing between low clouds and snow cover and in identifying cirrus clouds (Reynolds *et al.* 1978).

Platt *et al.* (1980) have carried out a careful study of cirrus cloud over Boulder, Colorado, as observed in the visible and infrared channels of the GOES-E and GOES-W geostationary satellites and as observed by ground-based lidar. Fig. 11.2 shows a plot of the albedo of the clouds versus their infrared emissivity compared with the predictions of radiative transfer calculations on two model

Fig. 11.2 Plot of cloud albedo versus cloud vertical emittance. Also shown are predicted curves for the two model clouds (see text) (from Platt *et al.* 1980).

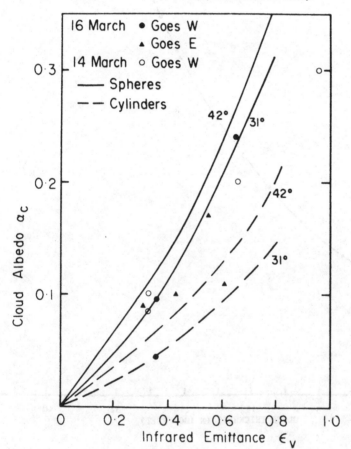

clouds, one assuming spherical particles and the other cylindrical particles. Agreement with the cylindrical particle model is good. Comparison with observations of back scattering from the ground based lidar demonstrated consistency between the lidar and satellite observations and the theoretical model. With the availability of two similar geostationary satellites with overlapping fields of view, namely GOES-E and GOES-W, the possibility exists of viewing clouds stereoscopically. Hasler (1981) has shown that cloud height, which we have seen is an important cloud parameter, can be determined to ±0.5 km or better, and that stereo observations can be applied to assist in determining motion from cloud tracking (para 12.2) and in studies of severe storms.

The effect of cloud on temperature sounding experiments is most serious when the clouds are thin, such as high level cirrus, because their effect can be large enough to introduce substantial errors into the measured profile but still too small to permit unambiguous detection of the cloud. Furthermore, clouds of this kind may not appear on photographs taken at visible wavelengths. Because the absorbing and scattering properties of ice crystals which make up this kind of cloud are strong functions of wavelength, a possible means for detecting cirrus is to make radiance measurements in spectral intervals chosen so that the transmission profile of the cloudless atmosphere is approximately the same for the different intervals, while the properties of ice crystals may be very different. The calculations of Houghton and Hunt (1971) show that a thin layer of cirrus, with a particle density of only 0.1 cm^{-3}, can produce a difference of 20 K between the brightness temperatures of the earth's atmosphere measured at 50 μm and 120 μm, where these would be the same in the absence of the cloud layer. Observations demonstrating this effect have been reported by Barnett *et al.* (1973).

McCleese (1978) has pointed out that detailed cloud properties may be derived by combining observations from a number of infrared channels. He analysed data from the channels included particularly for cloud study on the Selective Chopper Radiometer (SCR) on Nimbus 5 (§ 6.7). Data for four of the channels is shown in Fig. 11.3 for part of an orbit when the satellite passed over a particularly deep convective cloud at a latitude of about 15°S. As mentioned above, the difference between the effective brightness temperature in the two long-wavelength channels is particularly sensitive to cloud particle size (Houghton and Hunt 1971, Barnett *et al.* 1973) as is also the reflectivity of sunlight near 2.6 μm. The result of an attempt (McCleese 1978) to fit a cloud model including particle number density and particle size to the particular observations near 14°S of Fig. 11.3 is shown in Fig. 11.4. The solution shown may not be particularly unique, nor, of course, may the real cloud be of uniform structure over the instrument field of view of ~25 km diameter. However, McCleese points out that a thin layer of small particles is required to fit the 2.6 μm reflectivity and a thicker layer of large particles to fit the observations at the long wavelengths. Clearly very interesting information regarding cloud structure may be inferred by detailed comparison of a number of spectral channels.

Two techniques have been proposed for discrimination between ice and water clouds. One is to compare their near infrared albedo at wavelengths of 1.7 μm and 2.1 μm, the other is to make polarization measurements on reflected sunlight from them. The possibilities offered by exploiting polarization techniques have been stressed in the USSR and measurements between 0.3 μm and 3 μm made from the Meteor 8 satellite demonstrate their value.

Fig. 11.3 Radiances from four channels of the Nimbus 5 SCR for part of an orbit over the Indian ocean and Western Australia on the 15th January 1973 (from McCleese 1978).

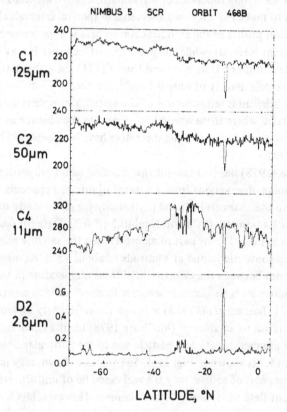

11.2   Aerosols

Aerosols or suspended particles in the atmosphere are important in two regions of the atmosphere. First, in the troposphere where, particularly over continents, they can interfere very substantially with the atmosphere's radiation budget; secondly in the stratosphere, where distinct layers exist with particles present in significant concentrations.

By using imaging photometers on satellites to observe solar radiation scattered back from the atmosphere, the presence of substantial dust clouds may often be inferred. Dust blown from the Sahara desert has been tracked substantial distances over the Atlantic ocean, for instance. Inferring quantitative information regarding the aerosol clouds is difficult, however, because of the problem of disentangling the effect of the aerosol from the variable background signals which arise due to the earth's surface and the remainder of the atmosphere. Proposals for utilizing measurements of the polarization of reflected radiation have been made as one method of identifying more clearly the effects of the aerosol.

A method of observing aerosols, particularly applicable to stratospheric particles, is to make twilight observations of the earth's limb. Such were reported from the Soyuz spacecraft of the USSR (Kondrat'yev *et al.* 1970).

Fig. 11.4 Retrieved cloud properties for cloud feature shown in Fig. 11.3 (from McCleese 1978).

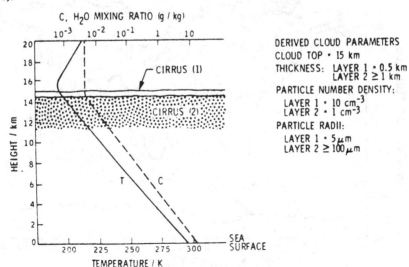

DERIVED CLOUD PARAMETERS
CLOUD TOP • 15 km
THICKNESS: LAYER 1 • 0.5 km
            LAYER 2 ≥ 1 km
PARTICLE NUMBER DENSITY:
  LAYER 1 • 10 cm^{-3}
  LAYER 2 • 1 cm^{-3}
PARTICLE RADII:
  LAYER 1 • 5 μm
  LAYER 2 ≥ 100 μm

Fig. 11.5 A sunset as viewed by the Nimbus-7 satellite. The instrument starts to scan the solar disk at a tangent height (h) of about 350 km, and follows it down until the sun disappears. Note the different layers of the atmosphere are successively sampled during the event (from McCormick *et al.* 1979).

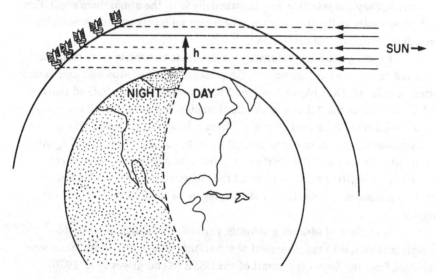

Fig. 11.6 SAGE sunrise and sunset tangent locations for February 1979. Successive events are shifted slightly in longitude and latitude. The + symbols represent sunsets and the • symbols represent sunrise (from McCormick *et al.* 1979).

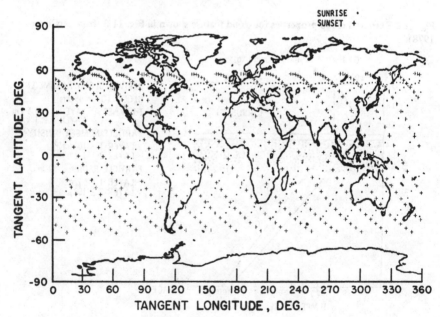

Recent instruments for making twilight observations of aerosol have been the Stratospheric Aerosol Measurement (SAM II) mounted on the Nimbus 7 satellite and the Stratospheric Aerosol and Gas Experiment (SAGE) mounted on a dedicated Applications Explorer Mission launched in February 1979. Both were built by a group at the Langley Research Centre, Virginia, in cooperation with the University of Wyoming (McCormick *et al.* 1979). Fig. 11.5 shows the geometry of the measurement appropriate to Nimbus 7 and Fig. 11.6 the coverage of sunset and sunrise locations for the SAGE for one particular month.

The SAM II instrument is a one spectral channel sun photometer (Fig. 11.7) with a passband centred at a wavelength of 1 μm in a region of the spectrum where absorption by atmospheric gases is negligible. Any attenuation of sunlight, therefore, is due to scattering by aerosol or Rayleigh scattering by molecules.

Fig. 11.7 The SAM II optical system (from McCormick *et al.* 1979).

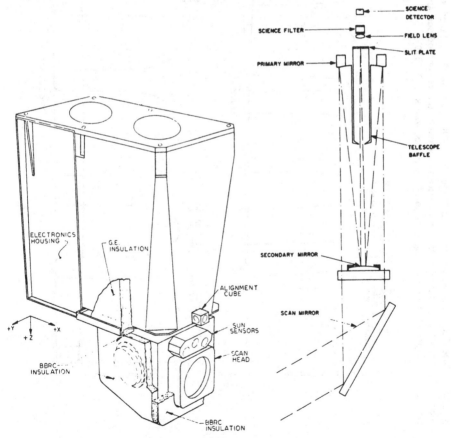

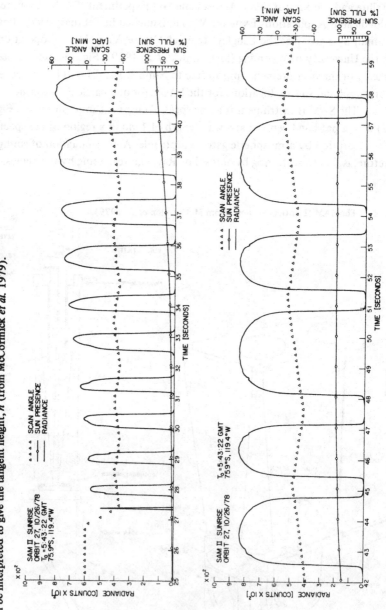

Fig. 11.8 A SAM II solar sunrise scan taken on 26 October 1978 during the 27th orbit of Nimbus 7. The latitude is 75.9°S. The ordinate (in counts) is directly proportional to the light intensity received by the pin photodiode, and the abscissa (in second from beginning of the scan) can be interpreted to give the tangent height, h (from McCormick et al. 1979).

During the sunrise or sunset period the instrument scans across the sun a number of times (Fig. 11.8). Note the flattening of the solar disc due to refraction near the beginning of the sequence.

The orbit of SAGE is arranged to complement the geographical coverage of SAM II on Nimbus 7. The SAGE instrument includes four channels at wavelengths of 0.385, 0.45, 0.60 and 1.0 μm respectively; the other channels are included to provide measurements of O_3 and NO_2 in addition to aerosol.

Aerosol profiles can be inverted by techniques described by Chu and McCormick (1979). Some results are given in McCormick *et al.* (1981, 1982). Fig. 11.9 shows enhanced aerosol extinction, especially in a layer near 20 km altitude, resulting from the eruption of Mt St Helens on 18 May 1980.

11.3 Precipitation

We have already seen that observations in the microwave part of the spectrum are particularly useful in making measurements of the water-vapour and liquid water content of the atmosphere over the oceans (§ 6.9). This is possible, because at wavelengths around 1 cm the emissivity of the ocean surface is low enough to present a cold background against which emission from water vapour and liquid water in the atmosphere may be observed. Measurements at a

Fig. 11.9 Aerosol extinction for volcanic enhanced conditons observed by the SAGE (after McCormick 1982).

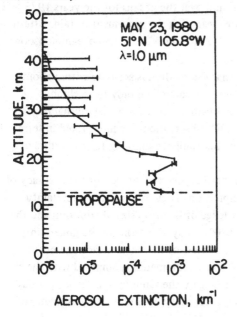

number of wavelengths assist in distinguishing effects due to water vapour from those due to liquid water; the varying emissivity of the surface, however, presents a problem in interpretation as has been pointed out by Rosenkranz *et al.* (1978).

Further the drop size is important in determining the interaction between microwave radiation and cloud particles. Clouds containing small droplets a few micrometres in size are substantially transparent to millimetre radiation. However, as the ratio of particle size to wavelength becomes comparable to unity, the absorption cross section of the droplets becomes appreciable. Early work on microwave sounding of precipitating clouds was carried out in the USSR by Basharinov *et al.* (1969) and Akvilonova *et al.* (1971).

Methods for the interpretation of microwave measurements near 1.5 cm wavelength in terms of precipitation rate have been developed by Wilheit *et al.* (1976 and 1977). They show that their estimates are not very sensitive to assumptions about the distribution of drop sizes but, because of the very different properties in the microwave region of liquid water and ice, their estimates are very sensitive to assumptions about the height of the freezing level. Wilheit *et al.* (1977) have compared rainfall estimates from the electrically scanned microwave radiometer (ESMR) on Nimbus 5 with estimates made from coincident observations from a radar based in Florida; they have also compared direct rainfall measurements with estimates made from observations using a ground-based upward-looking radiometer based at Goddard Space Flight Centre in Maryland. Fig. 11.10 shows the results of the comparisons. Using these methods Rao *et al.* (1976) have published an atlas of rainfall over the oceans for the years 1973 and 1974 which shows interesting features regarding the movement of rain patterns especially those associated with the ITCZ and with the Indian ocean monsoon circulation.

Wilheit *et al.* (1977) claim that the method is accurate within about a factor of two. ESMR observations, however, are taken only from Nimbus passes near to local noon and local midnight so, because of the diurnal rainfall cycle, uncertainty is introduced if ESMR observations are interpreted in terms of average rainfall rates. Further satellite instruments sampling at other times of day could overcome this problem.

Lovejoy and Austin (1980) have carried out a study of the accuracy of microwave remote sounding of precipitation. They estimate the uncertainty which arises because of lack of knowledge of the drop size distribution and the thickness of the rain layer as about ±70%. They also estimate the uncertainty which arises because of the large field of view of the satellite radiometers (of the order of 30 km for the Nimbus 6 and 7 radiometers) compared with the size of a typical rain cell as ~±200%. Sampling by the satellites only twice per day at given local times leads to a further uncertainty in rainfall, they estimate, of

±300%. Some improvement in accuracy can be expected as the field of view is reduced for future satellite instruments by employing larger antennae; a major problem to be overcome is that of providing for adequate sampling in time.

So far quantitative estimates of precipitation from microwave measurements have only been made over the oceans. Interpreting data over land is much more complex because of the variable background of the earth's surface. The prospects for overcoming this problem are discussed in a study of Nimbus 6 ESMR data by Rodgers *et al.* (1979).

A number of studies have been made employing other less direct methods of rainfall estimation from satellite observations. These are based on observations from visible or infrared imagery of cloud type or cloud brightness (see for instance Barnett 1974; Reynolds and Smith 1979). Garcia (1981) has compared two such techniques applied to observations over the Atlantic Ocean during the GATE experiment, with radar measurements at the same time, show-

Fig. 11.10 Brightness temperature as a function of rainfall rate. Nimbus 5 ESMR observations compared with radar measurements of rainfall rate (dots). The crosses are comparisons between ground based measurements of brightness temperature (converted into an ESMR brightness temperature scale) and direct measurements of rainfall rates at Goddard Space Flight Centre. The solid line is the calculated brightness temperature curve for a 4 km freezing level and the dashed lines represent a departure of 1 mm h^{-1} or a factor of 2 in rainfall rate (whichever is the greater) from the calculated curve (after Wilheit *et al* 1977).

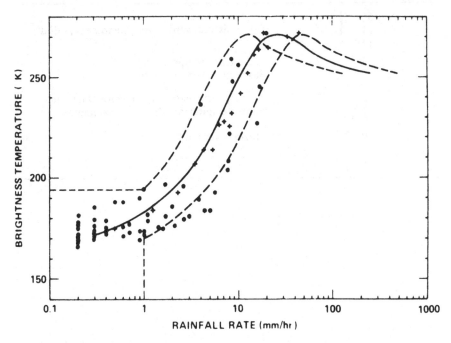

ing a useful level of agreement between the satellite and radar methods.

Once precipitation rates have been determined it is a relatively simple matter to derive patterns of latent heat release. Adler and Rodgers (1977) have carried out a study with ESMR data of the release of latent heat from a tropical cyclone and have shown that with such data it is possible to map the distribution of latent heat release as the storm develops and to compare the contributions to the latent heat release from different ranges of intensity of precipitation.

We have mentioned in a number of places measurements relevant to the atmosphere and the earth's surface which can in principle be made with passive microwave radiometry. It is convenient to list the possibilities at different wavelengths in the microwave spectrum. Table 11.1 (from Waters *et al.* 1975) lists the primary measurements which may be made at each wavelength. As we have seen, however, measurements at any wavelength are dependent on a number of parameters and the accuracy with which each can be determined from multispectral observations has yet to be established.

Table 11.1

Wavelength (cm)	Primary observable
50 21 11	Soil moisture, subsurface phenomena, salinity
4.6	Sea surface temperature
2.8	Sea state, heavy precipitation
1.5 1.3 0.9	Atmospheric water vapour, moderate and light precipitation, drop size parameter
0.57 0.26	Storms over land
0.32	Water-ice boundaries

12

DENSITY, PRESSURE AND WIND
MEASUREMENTS

12.1   Density and pressure

The most basic quantity required by the meteorologist is the detailed distribution of atmospheric density.

Density $\rho(z)$, pressure $p(z)$, composition and temperature $T(z)$ are related by the equation of state

$$p(z) = \frac{\rho(z)}{M(z)} RT(z) \qquad\qquad (12.1)$$

where $M(z)$ is the mean molecular weight as a function of height. Although the composition may change with regard to important minor constituents, these variations do not in the stratosphere and troposphere change $M(z)$ significantly, which may be regarded as constant and independent of time. Then the hydrostatic equation

$$dp(z) = -g\rho(z)\, dz \qquad\qquad (12.2)$$

may be used in the determination of pressure and density profiles ($p(z)$ and $\rho(z)$ respectively) if the temperature profile is measured and the pressure at some reference surface, $p(z)$, is known. The temperature profile on its own is not enough, although, because of the correlation between temperature and pressure fields, some success has been obtained with inferring $p(z)$ from temperature information alone.

One way in which satellites can help in the measurement of pressure at a reference surface is to use them as communication platforms to interrogate remote pressure-measuring devices mounted on free-floating buoys or on constant-density balloons free-floating in the lower stratosphere. Both buoys and balloons have been deployed during the Global Weather Experiment in 1979.

Measurements at millimetre wavelengths constitute a possible remote sounding technique. In § 5.1, measurements of atmospheric emission in the

0.5 cm oxygen band were described with the purpose of inferring atmospheric temperature structure. It is possible to employ the same band for measurements of the total mass of atmosphere above a given location (i.e. the atmospheric pressure) by measuring the absorption of an active microwave pulse emitted by a satellite transmitter, reflected from the surface and received again by the satellite (Smith *et al.* 1972). If adequate accuracy (better than 3 mbar, i.e. 0.3%) can be achieved with such measurements, they will be of great meteorological import-ance, especially over areas such as the oceans of the southern hemisphere where data are very scarce. Fig. 12.1 shows the spectral region involved and the absorp-tion by oxygen and water vapour. To eliminate the effects of water vapour, and

Fig. 12.1 Attenuation of a vertical path throughout the atmosphere due to oxygen and water vapour (2.5 g cm^{-2}) in the microwave region near 0.5 cm. A possible set of freq-uencies for the microwave pressure sounder are shown by triangles (after Peckham, private communication).

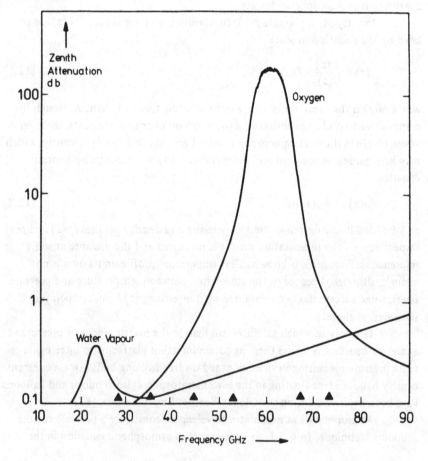

the varying reflectivity of the ocean surface, a number of spectral channels are proposed. Although it is known that thin non-precipitating clouds are sufficiently transparent not to affect the measurements, clouds containing large drops will have an effect. Since such clouds tend to be present in the areas where surface pressure measurements are particularly required it needs to be established just what coverage will be possible with this very promising measurement technique.

12.2 Wind measurement

At latitudes away from the tropics, >10° or so, the atmospheric wind field is related to the density field with reasonable accuracy through the use of the geostrophic approximation. As one approaches the equator, however, the Coriolis parameter becomes very small and the geostrophic approximation no longer applies. It is necessary, therefore, in tropical regions, to make direct wind measurements, and such measurements at higher latitudes will still provide important information about the density field.

Since cloud pictures have been available from geostationary satellites, considerable attention has been given to the deduction of winds from cloud motions. Suomi (1969) and Fujita *et al.* (1969) describe some of the first such investigations. Clouds at two levels are generally available for observation, cumulus clouds at levels around 900 mbar and cirrus clouds typically around the 200 mbar level. Details of the techniques of wind determination may be found in Hayden *et al.* (1979) and Bowen *et al.* (1979).

Two problems exist with regard to the interpretation of cloud motions in terms of winds. Firstly, do the levels being tracked move with the wind and secondly, how accurately can the cloud height be determined?

Regarding the first of these, Fujita *et al.* (1975) point out that substantial differences of up to ±40% occur between the wind velocities which would be deduced by tracking a single cumulus cloud in its growing stage or its dissipating stage. Hasler *et al.* (1977) have observed cloud motions from aircraft and compared them with wind velocities measured from the same aircraft. They find that for tropical oceanic cumulus, the average vector difference between the velocity at the cloud base and the wind velocity was 1.5 ms^{-1}, the average wind velocity being 8.5 ms^{-1}. The velocity at the cloud tops was, however, different from the 'wind' by a much larger amount of $\sim6 \text{ ms}^{-1}$, due presumably to the same effect as pointed out by Fujita *et al.* (1975). For cirrus clouds having a mean velocity of 11 ms^{-1} the mean vector difference between cloud and wind was 2 ms^{-1}.

When cloud images from satellites are being tracked, however, clouds are not observed singly but in groups so that the possible error arising from individual clouds not moving with the wind is much reduced. Suchman and Martin

(1976) studied the accuracy, representativeness and reproducibility of cloud tracer winds in the area of the GARP Atlantic Tropical Experiment (GATE). They found that different scientists using the same image data deduced winds differing by less than $2\,\text{ms}^{-1}$ at cirrus level and $1.3\,\text{ms}^{-1}$ at cumulus level, and that the vorticity and divergence patterns deduced from different analyses were similar. They also compared winds derived from satellite data with those derived from radiosondes launched from ships and found differences less than $3\,\text{ms}^{-1}$ which is close to the 'noise' level of the radiosonde determinations. They conclude that the derivation of wind from cloud tracers is as good, if not better, than that from the GATE network of radiosondes. Suchman and Martin admit that determination of cloud height with sufficient accuracy remains a significant problem.

Hubert (1976) and Davis *et al.* (1976) also report studies comparing winds derived from satellite images with conventional wind data. They find somewhat larger differences than those reported by Suchman and Martin (1976) and also emphasize particularly the problem of height determination especially for cirrus clouds. Heights are determined by observing the cloud in the infrared, thereby determining a radiometric temperature. If an emissivity for the cloud and a profile of temperature against height is then assumed, a height can be assigned to the cloud tops. The largest uncertainty here (especially with cirrus clouds) is in estimating the cloud emissivity, which may vary over a wide range. The experience of the operator can be an important factor in assigning a value for cloud emissivity from considerations such as cloud type and synoptic situation. Stereoscopy (Hasler 1981) can also be a valuable aid to the estimation of cloud altitudes.

The usefulness of satellite-derived winds is clear from the many studies in the literature which make use of them. For example, a climatology of low-level winds from satellite images has been published by Gaby and Poteat (1973), and studies of the flow around a particular convection cell have been carried out by Sikdar and Suomi (1972) and by Suchman *et al.* (1977). These latter studies demonstrate that realistic divergence fields can be produced from satellite-derived winds from which estimates of vertical velocities can be made giving values in reasonable agreement with estimates from other considerations. From covariance analyses of satellite-derived wind fields, Maddox and Vonder Haar (1979) estimate that the random error in vector wind fields derived from cumulus cloud tracking is less than $1.75\,\text{ms}^{-1}$.

In Fig. 2.8 is shown the coverage by the five geostationary satellites which made observations during 1979 – the year of the First GARP Global Experiment – and from which tropical wind fields can be determined.

Late in 1977 the European Geostationary Satellite, METEOSAT, was launched (§ 2.2). The radiometer on board included for the first time in geo-stationary orbit a 6.3 μm water-vapour channel which responds to emission from water vapour in the upper troposphere (Fig. 12.2). In regions of high radiance in the 6.3 μm channel the emission originates from relatively low levels in the atmosphere; these regions are, therefore, relatively dry in the upper troposphere. Regions of low radiance, by contrast, possess high upper tropospheric humidities. From film loops of successive images which have been completed at the European Satellite Operations Centre at Darmstadt, it is clear that movements of dry and wet areas can be rather clearly seen in the 6.3 μm data. That data from this band can be used for wind determination has, in fact, been demonstrated by

Fig. 12.2 Image from the 6.3μm channel on Meteosat for 8 February 1978 at 1155 Z showing distribution of upper tropospheric water vapour.

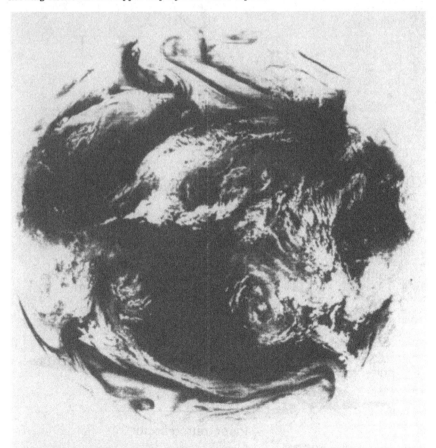

METEOSAT 1978 MONTH 2 DAY 8 TIME 1155 GMT (NORTH) CH. WV NOMINAL SCAN/PROCESSED SLOT 24 CATALOGUE 1000220036

Steranka *et al.* (1973) from Nimbus 4 data and by Kastner *et al.* (1980) from Nimbus 5 data. Meteosat data clearly can be employed in a similar way.

It was mentioned in § 3.3 that the emissivity of the sea-surface is dependent on sea state and is, therefore, related to wind speed. Measurements from the microwave radiometers SMMR on Nimbus 7 and on SEASAT have been interpreted in terms of surface wind with some success (Lipes *et al.* 1979). The shape of the return signal from the radar altimeter on SEASAT is also dependent on wave height and hence is related to wind speed, although there is no information from the altimeter regarding wind direction (Tapley *et al.* 1979).

Fig. 12.3 The Seasat satellite and its instruments (after Williams and McCandless 1976).

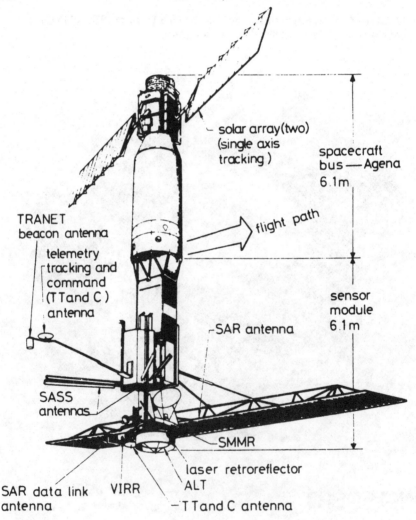

If instead of viewing vertically downwards, a sidewards looking radar is employed, both wind speed and direction may be inferred from the intensity of the backscatter from the sea surface. A suitable radar known as the SEASAT-A Scatterometer System (SASS) was mounted on the Seasat satellite launched in 1978 (Jones *et al.* 1979). Bragg scattering of the microwaves occurs from centimetre size capillary ocean waves. The strength of the backscattered signal is proportional to the capillary wave amplitude which is proportional to the wind speed near the sea surface. Moreover because the radar backscatter is anisotropic the wind direction can be derived from SASS measurements at different azimuths.

The SASS radar operated at a frequency of 14.6 Hz with 100 W peak RF power. It illuminated the sea surface with four fan-shaped beams from four dualpolarized (vertical and horizontal) antennae (shown in Fig. 12.3). The antennae beams were oriented at 45° relative to the subsatellite track yielding observations separated in azimuth by 90°. Because of the satellite's velocity the returned signals were shifted in frequency so that with the aid of 12 Doppler filters the antennae footprint could be separated into resolution cells approximately 50 km on a side. From a programme of aircraft observations associated with surface wind observations empirical relations were derived relating the strength of the backscattered signals to the wind vector as a function of incidence angle, azimuth angle and polarization. Using these relations, wind vectors derived from SASS were compared with ships and buoys in the Gulf of Alaska. Results shown in Fig. 12.4 for the satellite-buoy comparison indicate that SASS derived winds are accurate to $\pm 2\,\text{ms}^{-1}$ in speed and $\pm 20°$ in direction – in fact they have an accuracy equal to or greater than conventional surface observations.

In the stratosphere and mesosphere the possibility has been proposed of wind measurements by observing the Doppler shift of suitable emission lines from the atmospheric limb (Rees 1980). Another possibility for observing the Doppler shift is by employing a gas correlation radiometer (Drummond *et al.* 1980) observing nearly perpendicular to the spacecraft's velocity. To obtain a useful accuracy in wind measurement, however, makes very stringent demands on the spacecraft attitude control and stability, since the contribution of the component of the spacecraft's velocity along its orbit to the line of sight velocity vector must be known to high precision.

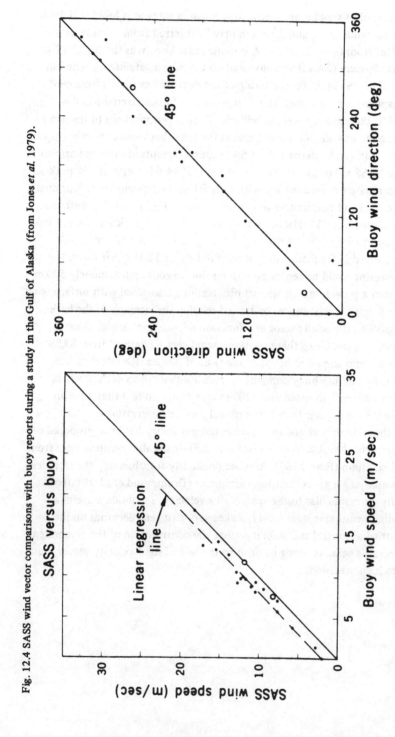

Fig. 12.4 SASS wind vector comparisons with buoy reports during a study in the Gulf of Alaska (from Jones et al. 1979).

13

IONOSPHERIC MEASUREMENTS

13.1 Introduction

Remote sounding of the ionosphere from observations of the propagation of radio waves has been pursued vigorously for forty years or more and was one of the first applications of remote sounding techniques to the atmosphere. A single ionosonde station on the ground enables the vertical distribution of electron density up to the level of peak density to be obtained as a function of time. One of the first applications of satellites in this field was to use transmission measurements (especially Faraday rotation of the plane of polarization) of the region between a satellite and a ground station by observing the behaviour of the received signal from the satellite beacon as the range of the satellite varies; from such observations the total electron content of the region can be determined.

The ability of a polar orbiting satellite to obtain global coverage has been exploited in the topside sounder, a satellite-mounted instrument for probing the top part of the ionosphere in the same way as a ground-based ionosonde probes the lower part. A number of such satellites have been launched and have enabled for the first time the global distribution of electron density in the F-layer to be measured.

13.2 Theory of radio probing

The theory of the propagation of radio waves in an ionized medium is well known (*cf.* Ratcliffe 1959, Budden 1961), and will, therefore, only be summarized here.

In a region where the number density of electrons of charge e and mass m is N m^{-3}, and the magnetic field is B, two frequencies are important, namely, the plasma frequency

$$\omega_N{}^2 = \frac{Ne^2}{m\epsilon_0} \tag{13.1}$$

and the gyrofrequency $\omega_H = Be/m$. For $f_N = 10^6$ Hz, $N = 1.2 \times 10^{10}$ m^{-3}. Let the frequency of incident radio waves be ω, and $X = \omega_N^2/\omega^2$, $Y = \omega_H/\omega$, $Y_L = Y \cos \theta$, $Y_T = Y \sin \theta$ where θ is the angle between the direction of the wave normal and the magnetic field. In a region of negligible absorption, the Appleton-Hartree equation for the refractive index is

$$\mu^2 = 1 - \frac{X(1 - X)}{(1 - X) - \frac{1}{2}Y_T^2 \pm \{\frac{1}{4}Y_T^4 + (1 - X)^2 Y_L^2\}^{1/2}} \tag{13.2}$$

where the + sign refers to the ordinary (O wave) mode and the – sign to the extraordinary modes (X or Z waves). The lowest frequencies for which waves of a particular mode (Z, O and X) can propagate are called the cut-off frequencies. Fig. 13.1 shows an ionogram taken from the Alouette II satellite showing reflections from the ionized layer, cut-offs and also various resonance spikes. The electron density at satellite height is immediately deduced from the cut-off frequencies at zero range. The electron density distribution through the top side of the ionosphere can be found from the curve of apparent range h' versus frequency which is described by the equation

$$h' = \int_0^h \mu_g(\omega, N)\, dh \tag{13.3}$$

where h is the true range and μ_g the group refractive index. Various numerical techniques for inverting equation (13.3) have been devised.

Further information can be obtained from the observation of a number of resonances which arise from combinations of the f_N and f_H frequencies, and from the observation of echoes reflected from the earth's surface at frequencies greater than the cut-off frequency.

13.3   Instrumentation and results

Five satellites for topside sounding of the ionosphere have been launched to date, Explorer XX (1964), Alouette I (1962), Alouette II (1965), ISIS I (1969) and ISIS II (1971). The Alouette and ISIS satellites have been part of a joint US-Canadian programme. Explorer XX carried a set of fixed frequencies only for sounding. The Alouette satellites carried swept frequency sounders and the ISIS satellites carried both. Fig. 13.2 shows a sample of electron density information deduced from such satellites.

Images of the aurora taken from satellites, first from ISIS II, from a number of the US Defence Meteorological Spacecraft and most recently from the Dynamics Explorer launched in 1981 (Frank *et al.* 1981), have shown for the first time the overall morphology of the aurora and how its features change during magnetic storms (Fig. 13.3).

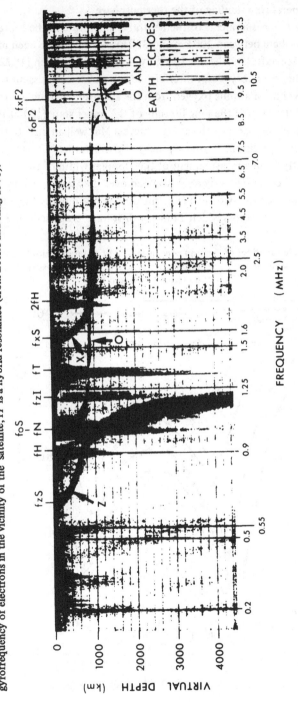

Fig. 13.1 Alouette II ionogram acquired at 2212 UT on 29 July 1967 when the satellite was above a point 30°S 161°E. Low frequency cut-offs for the three modes of propagation (O, X and Z) are labelled fOS, fXS, fZS and correspond to the frequencies at which waves are reflected at the satellite height. Critical frequencies for the F₂ layer are labelled fOF₂ and fXF₂ fZI is the high frequency cut-off of the Z wave. Resonances are also shown: fN, the plasma resonance coinciding with fOS; fH, the electron gyro resonance when the transmitter frequency is equal to the gyrofrequency of electrons in the vicinity of the satellite, fT is a hybrid resonance (from Eccles and King 1970).

13.4   Temperature sounding of the thermosphere

Remote measurements of the neutral temperature in the F region part
of the thermosphere between 200 km and 320 km altitude have been made from
the OGO 6 spacecraft, launched in 1969, by Blamont and Luton (1972). They
mounted a spherical Fabry–Perôt interferometer to observe emission from the
6300 Å airglow line of atomic oxygen from the atmospheric limb as viewed from
the spacecraft. Because of the long lifetime of the upper state (^{1}D) of the transition, oxygen atoms raised to this level achieve a Maxwellian distribution of
velocity by elastic collisions so that the neutral temperature in this region can be
deduced directly from a measurement of the Doppler width of the 6300 Å line.

The spherical Fabry–Perôt is scanned over the airglow line by varying
the distance between the two elements, one of which is mounted in a tube of
piezoelectric material across which a saw tooth voltage is applied, thus generating

Fig. 13.2 Latitude variations of electron concentrations at different altitudes along the
110° E meridian deduced from observations from Alouette I which crossed the geographic
equator at 1203 LMT on 19 June 1963 (from Eccles and King 1969).

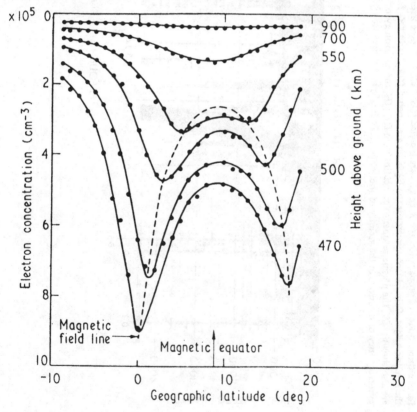

a continuous scan. The altitude being observed can be chosen by moving a mirror which alters the direction of view of the instrument. Constant adjustment of this direction is in any case necessary because of the alteration in altitude of the OGO spacecraft. Temperatures have been measured with an accuracy of ±65 K. The results demonstrate the effect of magnetic storms on the temperature in the polar regions. At 260 km altitude a temperature rise of 300 K has been noted during such a storm.

A more advanced Fabry–Perôt interferometer (Hays *et al.* 1981) has been built for the Dynamics Explorer (DE) satellite launched in 1981. Shown in Fig. 13.4, it contains an etalon, developed by Rees (1982), of spacing 1.26 cm stable to 10^{-7} cm, which generates an interference pattern which is recorded by an imaging photo cathode in the image plane. Atomic oxygen lines at 6300 Å (^{1}D), 7320 Å (0^{+2}P) and 5577 Å (^{1}S), the ^{2}D line of atomic nitrogen at 5200 Å and the ^{2}P line of sodium at 5896 Å can all be observed; emission from these lines spans the altitude range 90 km to about 250 km. From the line width, temperature can be deduced and from the line shifts the wind velocity along the line of sight. Killeen *et al.* (1982) have compared wind measurements from the Fabry–Perôt interferometer with *in situ* measurements from other instruments on the DE spacecraft.

Fig. 13.3 Schematic diagram showing the main characteristics of auroral displays. Discrete arcs are indicated by lines and diffuse auroral regions shaded (from Akasofu 1976).

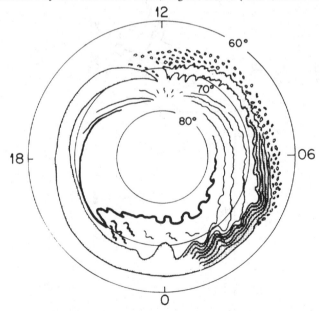

238

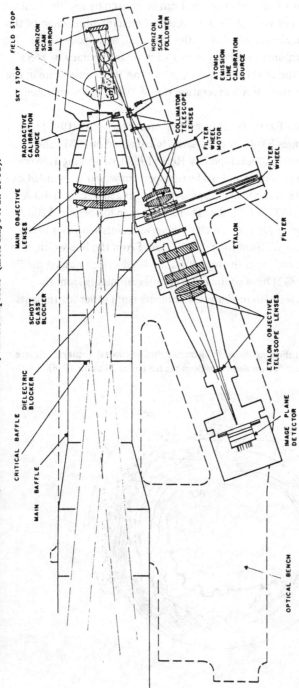

Fig. 13.4 Optical schematic for the Fabry-Perôt interferometer on Dynamics Explorer (after Hays *et al.* 1981).

14

REMOTE SOUNDING OF THE ATMOSPHERES OF VENUS AND MERCURY

Mercury is the planet closest to the Sun. It was long suspected that Mercury must have, at most, an exceedingly tenuous atmosphere, because the intense solar heating would allow most gases to escape the small planet's weak gravitational field. In 1974, instruments on board Mariner 10 confirmed this expectation. In fact, the atmosphere of Mercury was even more tenuous than might have been expected; the total surface pressure is less than 2×10^{-9} atmospheres, where as much as 10^{-6} atmospheres might have been anticipated (Kumar, 1976). Furthermore, the solar occultation experiment carried out by the ultraviolet spectrometer experiment detected no atmospheric signature at all. The UV instrument is shown in Fig. 14.1 (Broadfoot *et al.* 1977). It is an objective grating device, covering the range 0.02 to 0.17 μm with a resolution of 0.002 μm (20 Å). Channel electron multiplier detectors are used.

In its limb-viewing mode, the UVS had higher sensitivity to individual species and was able to detect a tenuous helium atmosphere, about 2×10^{-12} atmospheres (Broadfoot *et al.* 1974). Traces of atomic hydrogen were also detected (Broadfoot and Shemansky 1975), but no other species. The presence of helium is explicable either by solar wind accretion, or by the radioactive decay of uranium and thorium in the crust. The absence of volatiles such as H_2O and CO_2 indicates an inactive crust, or perhaps a deficiency of these materials throughout the planet (Kumar 1976).

Venus has been intensely studied because it is so close to the Earth, and our planet's twin in size and mass. Venus has an extremely hot, dense and active atmosphere, and exhibits nearly unbroken cloud cover. The portion above the clouds has mean temperature and pressures close to those found in the Earth's atmosphere and appears to have analogous Sun-driven meteorological behaviour. The deep atmosphere, with pressures up to 100 bars and temperatures as high as 750 K, has radiative time constants too long to respond to the diurnal cycle and is more analogous to the terrestrial oceans. The upper atmosphere is therefore a

240

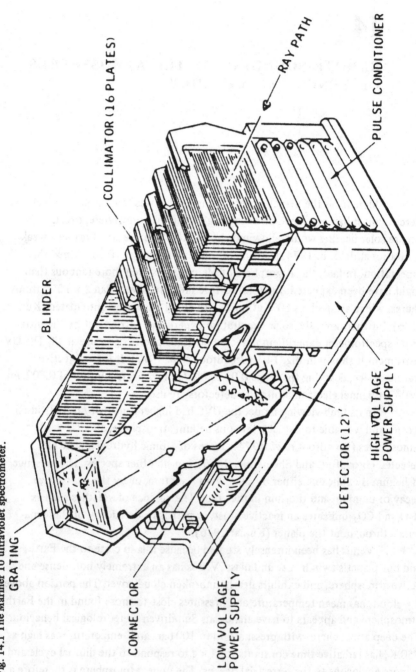

Fig. 14.1 The Mariner 10 ultraviolet spectrometer.

RAY PATH

COLLIMATOR (16 PLATES)

PULSE CONDITIONER

BLINDER

GRATING

DETECTOR (12)'

HIGH VOLTAGE POWER SUPPLY

CONNECTOR

LOW VOLTAGE POWER SUPPLY

more fruitful target for remote investigations. Such measurements were made from Mariners 2, 5 and 10, Veneras 9 and 10, and the Pioneer Venus Orbiter.

For atmospheric investigations, Mariner 2 carried both microwave and infrared radiometers, arranged to scan the same regions of the planet simultaneously. The infrared radiometer is shown schematically in Fig. 14.2. Immersed thermistor bolometer detectors were used with a chopping speed of 20 Hz (Chase *et al.* 1963). The two channels had band-passes of approximately 8–9 μm and 10–10.8 μm; the former is a region of relative transparency for the Venus atmosphere, which consists largely of carbon dioxide, while the latter encloses the 10.3 μm band of that molecule. The 10.3 μm band is sufficiently weak that, in the presence of unbroken cloud cover, both channels sensed almost the same temperature, equal to that at the cloud tops. Regions of low or broken cloud, however, would allow the 8.5 μm channel to penetrate deeper than the 10.3 μm channel, sensing radiation from deeper, warmer levels. In this way it was hoped that structural details such as 'holes' in the clouds would be revealed. The radiometer scanned the planet with a square 1.2° field of view, corresponding to a resolution of about 5% of the disc of the planet at closest approach; no cloud structure details revealed themselves except an anomaly in the Southern Hemisphere which was 10 K cooler than its surroundings, indicative of a high cloud region. The mean cloud-top temperature was measured to be about 240 K on

Fig. 14.2 Schematic layout of Mariner 2 infrared radiometer (from Chase *et al.* 1963).

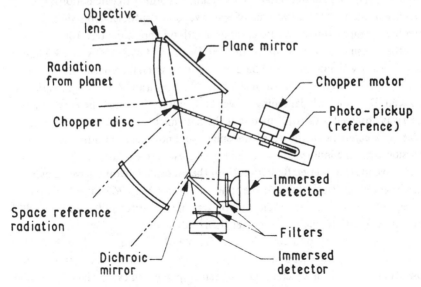

both the day and night sides. However, it is important to remember that Mariner 2, like all spacecraft to visit Venus until Pioneer in 1978, travelled in a plane very close to the ecliptic and hence did not observe the polar regions. The axis of rotation of Venus is within a few degrees of being perpendicular to the ecliptic plane, and so remote sensing of high latitudes on Venus is not possible using earth-based telescopes either. The discovery of complex cloud morphology on the planet near the pole was therefore left for the Pioneer polar orbiter.

The Mariner 2 microwave radiometer (Barath *et al.* 1964) consisted of an aluminium dish 48.5 cm in diameter with a field of view of 4.4° in a 13.5 mm wavelength channel and 5° in a 19 mm channel; both frequencies were measured simultaneously using a diplexer and separate video detectors. Calibration was achieved using two horn-type antennas which pointed continuously at cold space, and a noise discharge tube producing a standard signal equivalent to about 350 K. Since the clouds are substantially transparent to microwave frequencies, soundings of the deep atmosphere and surface are obtained. The principal sources of opacity at these wavelengths are the pressure-induced pure rotational transitions of carbon dioxide. For the amount of CO_2 in a vertical column on Venus, the opacity between the spacecraft and the surface is large at wavelengths short of a few mm, declining to small values longward of a few cm. One of the pure rotational lines of water vapour lies at 13.5 μm wavelength, and in the absence of water in the deep atmosphere of Venus the two Mariner 2 microwave channels were expected to measure brightness temperatures of about 550 K at 19 mm and about 430 K at 13.5 mm. According to calculations, the presence of about 0.5% by volume of water vapour is sufficient to reduce the brightness temperature at the longer wavelength by approximately 100 K. In fact, the measurements were 493–606 K at 19 mm and 393–400 K at 13.5 mm, favouring a dry Venusian atmosphere but with considerable uncertainty due to noise and other sources of error in the data. No significant day to night variation was seen. Definite limb darkening (lower temperatures towards the edge of the disk) was observed; this confirmed that the high microwave temperatures were indeed characteristic of the deep atmosphere and not due to non-thermal emission from the ionosphere, as had been suggested to explain Earth-based microwave measurements. Radiation from the ionosphere would have shown limb brightening. Thus, it was at the time of Mariner 2 in 1962 that the popular concept of Venus as a wet, tropical world was finally ousted in favour of a hellish environment with temperatures hotter than boiling oil.

Mariner 5 encountered Venus in October 1967 and passed behind the planet, permitting the first studies of the atmosphere by radio occultation. This powerful remote-sensing technique has been employed on every planetary probe since then, and some explanation of the principles involved, and the complexity of the data reduction operation, is appropriate at this point.

Mariner 5 performed two independent radio propagation experiments, one using the S-band (2.3 GHz) carrier frequency of the communications system, originating in the spacecraft, and the other using signals beamed from Earth and received by phase-locked receivers in the spacecraft. Just before encounter with Venus, the propagation time for the S-band signal transmitted from Mariner 5 to Earth was 4 minutes and 26 seconds; as the spacecraft passed behind the planet both the velocity and direction of propagation of the radio waves are modified by their passage through the atmosphere. The frequency, phase and strength of the signal are recorded and corrected for effects due to the terrestrial atmosphere and ionosphere. Then, provided the precise position and velocity of the spacecraft are known, the measurements are interpreted in terms of profiles of temperature and pressure *vs* height, as follows.

The expression for the frequency shift due to the presence of the atmosphere of Venus is (Fjeldbo *et al.* 1971).

$$\Delta f = (f_s - f_s \frac{V_{rs}}{c} \cos(\beta_c - \beta_r) - f_s \frac{V_{zs}}{c} \sin(\beta_c - \beta_r)$$

$$+ f_s \frac{V_{rt}}{c} \sin(\delta_s - \delta_r) + f_s \frac{V_{zt}}{c} \cos(\delta_s - \delta_r)) - (f_s - f_s \frac{V_{rs}}{c} \cos\beta_e$$

$$- f_s \frac{V_{zs}}{c} \sin\beta_e + f_s \frac{V_{rt}}{c} \sin\delta_s + f_s \frac{V_{zt}}{c} \cos\delta_s) \qquad (14.1)$$

where

f_s is the spacecraft transmitter frequency,

V_{rs} is the radial velocity of the spacecraft,

V_{zs} is the velocity of the spacecraft in the z direction,

V_{rt} is the radial velocity of the tracking station,

V_{zt} is the velocity of the tracking station in the z direction,

and the angles and coordinate system are defined in Fig. 14.3. A second equation

$$- z_t \sin(\delta_s - \delta_r) = (r_s^2 - z_s^2)^{1/2} \sin(\beta_c - \gamma - \beta_r) \qquad (14.2)$$

results if spherical symmetry is invoked. Fjeldbo *et al.* (1971) describe how these equations may be solved iteratively for the ray path angles δ_r and β_r. Then the asymptotic distance a of the ray path from the centre of mass of the planet and the angle of refraction α are calculated from

$$\alpha = \delta_r + \beta_r$$

$$a = (r_s^2 + z_s)^{1/2} \sin(\beta_c - \beta_r - \gamma). \qquad (14.3)$$

Successive measurements of rays which traverse increasingly deep levels in the atmosphere result in a profile of $\alpha(a)$; if the radius of the planet is known this can be converted into a profile of $n(z)$, where n is the refractive index of the atmosphere at altitude z. To proceed further, a composition for the atmosphere must be assumed. By the time of Mariner 5, the Russian entry probes Venera 4 and 5 had measured CO_2 mixing ratios of 0.9 ± 0.1 and 0.97^{+3}_{-4} respectively, the remainder thought to be mostly N_2 with small amounts of CO, HCl, HF and H_2O. The relationship between refractive index and density for CO_2 and N_2 are known from laboratory measurements and so a profile of molecular number density can be formed. Finally, this is interpreted in terms of the temperature and pressure profile using the hydrostatic equation and the equation of state. It is necessary to fix the temperature at the top of the recovered profile independently, as a boundary condition, and this, together with uncertainties in the composition and in the frequency stability of the transmitter, is the principal source of error. Fig. 14.4 shows the temperature profile on the night side of Venus, inferred from the S-band occultation. The divergent profiles at high altitudes result from different assumptions of boundary temperature, showing how the importance of this choice decreases with decreasing altitude.

Fig. 14.3 Geometry of the occultation experiment (from Fjeldbo *et al.* 1971).

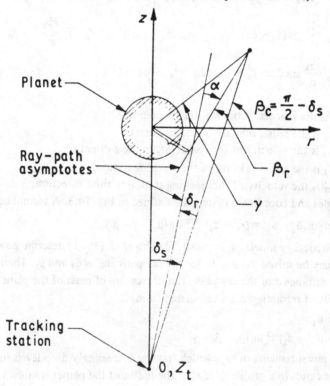

The independent 'up link' experiment used transmissions at two harmonically related frequencies, 49.8 MHz and 423.3 MHz. These were observed in frequency shift and amplitude every 0.6 s by a receiver on the spacecraft. The use of Earth-based oscillators resulted in better frequency stability than the spacecraft transmitter, and the two frequencies experienced different levels of attenuation at given atmospheric levels and hence permitted acquisition of data over a wider range of heights than either frequency used singly. An important source of noise in this type of measurement is due to scintillations caused by irregularities in the Earth's ionosphere; in the case of the Mariner 5 uplink experiment at the lower frequency, these were serious enough to render the data uninterpretable. However, the 423.3 MHz data were successfully reduced and led to profiles which were not significantly different from those obtained by the S-band experiment. Both day and night temperature and pressure profiles were obtained, but little diurnal effect was indicated.

The radio occultation experiment on Mariner 5 also yielded information on the ionosphere of Venus (Mariner Stanford Group 1967). The principles of this kind of measurement are discussed in Ch. 15. Both day and night-time ionospheres were found, with the night-time peak concentration of electrons about 1% of the day-time maximum, but both are thin by terrestrial standards

Fig. 14.4 Temperature profile on the night side of Venus. The inset shows the S-band loss coefficient (in db km^{-1}). From Rasool and Stewart (1971).

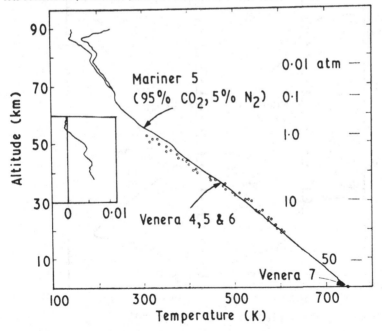

(Fig. 14.5). The Mariner 5 ultraviolet photometer (Barth *et al.* 1967) had three channels (1050-2200 Å, 1250-2200 Å, and 1350-2200 Å) using filters of lithium, calcium and barium fluorides on individual photomultiplier tube detectors with fields of view of 1.2°, 3° and 3° respectively. The Lyman-α airglow of atomic hydrogen was observed in the outer atmosphere, indicating an abundance of H comparable to that on Earth, but in a much thinner layer. This, together with the absence of detectable atomic oxygen emissions, implies a much lower exospheric temperature on Venus than on earth.

Remote sensing of the atmosphere of Venus continued in February, 1974 with the flyby of Mariner 10 on its way to Mercury. The imaging system on this spacecraft consisted of a television camera (Fig. 14.6) equipped with various filters (Table 14.1) mounted on a rotating filter wheel. Most significant among these for atmospheric investigations was the ultraviolet filter, for it had been known since 1926 that Venus exhibits faint markings in the ultraviolet which are absent at visible wavelengths (Fig. 14.7). The Mariner 10 images

Fig. 14.5 Profiles of electron density N_e in the ionosphere of Venus, obtained from the Mariner 5 dual frequency experiment. The vertical lines indicate the upper limits on ion density N_i measured by the Venera 3 entry probe. A typical terrestrial ionospheric profile is also shown (from Grinqauz and Breus 1970).

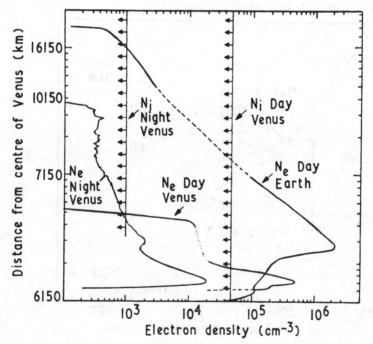

Fig. 14.6 Schematic view of Mariner 10 television camera (from Murray et al. 1974).

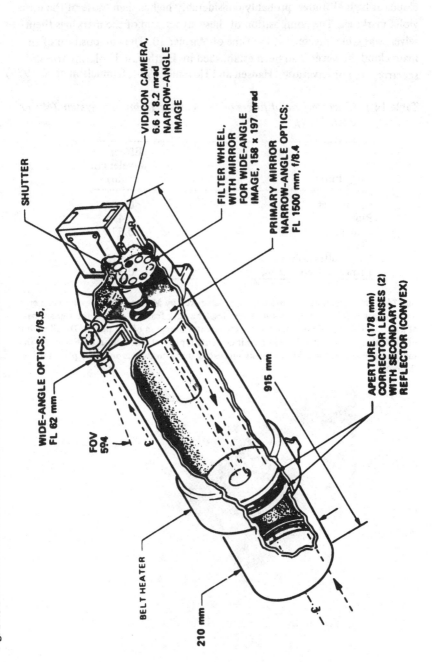

SHUTTER

VIDICON CAMERA,
6.6 × 8.2 mrad
NARROW-ANGLE
IMAGE

FILTER WHEEL,
WITH MIRROR
FOR WIDE-ANGLE
IMAGE, 158 × 197 mrad

PRIMARY MIRROR
NARROW-ANGLE OPTICS:
FL 1500 mm, f/8.4

WIDE-ANGLE OPTICS: f/8.5,
FL 62 mm

FOV
594

915 mm

BELT HEATER

210 mm

APERTURE (178 mm)
CORRECTOR LENSES (2)
WITH SECONDARY
REFLECTOR (CONVEX)

revealed these in extraordinary detail, as structure in the ubiquitous cloud veil (Murray *et al.* 1974). Observations near the limb show definite layers of thin clouds at high altitudes, probably considerably higher than those of the ultraviolet markings. The composition of these layers, and of the markings themselves, was still a mystery at the time of Mariner 10. The composition of the main cloud, however, had been established by Earth-based polarimetric and spectroscopic observations (Hansen and Hovenier 1974; Samuelson *et al.* 1975)

Table 14.1   *Characteristic of filters of the Mariner 10 imaging system (Murray et al. 1974).*

Filter	Effective wavelength (μm)
Ultraviolet	0.355
Blue	0.474
Orange	0.578
Clear	0.482
Minus ultraviolet	0.512
Ultraviolet polarizing	0.358

Fig. 14.7 This ultraviolet picture of Venus was taken by Mariner 10's television cameras on February 6 from a range of 490 000 miles. The dark features toward the top are part of a dark belt in the Venus clouds over the equatorial region of the planet. Detail within this belt shows rising and descending air currents typical of convection on Earth. To the south of the belt are spiral-like streaks suggesting uniform flow around the planet toward the pole.

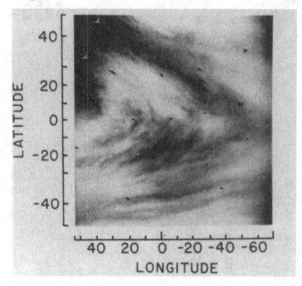

to consist of concentrated sulphuric acid solution in spherical droplets about 1 μm in radius.

Successive Mariner 10 images of Venus over an 8-day period allowed measurements of the speed and direction of motion of the clouds. Equatorial features travel completely around the planet in 4 to 5 days, corresponding to a zonal velocity of approximately 100 ms^{-1}, some 80 times faster than the rate of rotation of the surface of the planet. Meridional velocities are much smaller, less than about 10 ms^{-1} (Suomi 1975). Evidence for waves, turbulence and convection cells is also apparent (Belton *et al.* 1976). Thus, Venus is revealed as having an active and complex meteorology in its upper layers, where conditions of temperature and pressure are similar to those near the surface on Earth.

Mariner 10 also carried an infrared radiometer and an ultraviolet spectrometer. The radiometer (Chase *et al.* 1974) was a simple two-channel device with 8-14 μm and 35-55 μm spectral bands, designed principally for observations of temperatures on the surface of Mercury. The measurements of Venus obtained temperatures of 227 K for the shorter and 250 K for the longer wavelength channel, at the minimum zenith angle (viewing angle relative to the vertical) of 36.1°. Perhaps the most interesting feature of the Mariner 10 IR observations was the obtaining of the first good limb-darkening curves for Venus. These are not obtainable from Earth because of the high spatial resolution needed to obtain large zenith angle coverage. Interpretation of the simultaneous limb-darkening curves at two wavelengths suggest (Taylor 1975) that the opacity in the higher part of the region of the atmosphere which was covered is increasing rather than decreasing with height. Two possible explanations were layering in the main cloud deck, or water vapour in the atmosphere above the clouds which is depleted at lower levels due to the drying action of the sulphuric acid clouds.

The ultraviolet spectrometer (Broadfoot *et al.* 1977) was described above in connection with its measurements at Mercury. Table 14.2 shows a summary of the results for Venus, including the first positive detection of helium in the Venus atmosphere. The Lyman-α intensity observed by Mariner 10 was at least a factor 2 higher than that from Mariner 5, apparently the result of fluctuations in the concentrations of hydrogen and/or deuterium in the exosphere (1000-3000 km altitude). If the observed intensity was attributed entirely to resonance scattering by H atoms, then a mean exospheric temperature of 400 K was deduced. The escape rate for atomic hydrogen at this temperature is only 10^4 atoms cm^{-2} s^{-1}, four orders of magnitude less than the corresponding rate at Earth and Mars (Broadfoot *et al.* 1974).

The radio occultation experiment on Mariner 10 was of the dual frequency type, having X-band (8.4 GHz) for the first time. The system is shown schematically in Fig. 14.8. When the spacecraft passed behind the planet, the

X-band signal was completely lost at an altitude of 51 km above the surface, while the S-band signal persisted down to about 40 km before fading out. An analysis of the S-band entry data in terms of vertical temperature profile showed the presence of four distinct temperature inversions. It has been suggested that these are associated with layers of different stability in the atmosphere, or perhaps with latent heat absorption or release caused by condensation or evaporation of cloud layers.

The electron density profiles on the day and night sides of Venus as measured by Mariner 10 have been presented by Fjeldbo *et al.* (1975). They show a gross similarity with the Mariner 5 results, but the day side levels fall off much faster at higher levels, and the night-side profile shows a distinct two-layered structure. In 1975, the USSR achieved the first Venus orbiters with Veneras 9 and 10, which also delivered landers to the surface. Each orbiter had a full complement of remote sensing experiments, as detailed in Table 14.3. Many of these instruments were similar to those on the 1973 missions to Mars. All of the optical instruments (except the TV) have fields of view of about one degree,

Table 14.2 *Comparison of dayglow observed at Venus and the Earth by Mariner 10. The data for the Earth were obtained at 2130 G.M.T., 3 November 1973, at 282 000 km, with a cone angle of 85.3°. Data for Venus were obtained on 5 February 1974 at two distances; the conditions were: 194 600 km, 2315 G.M.T., cone angle 151.8°; and 13 000 km, 1710 G.M.T., cone angle 131.0°. The numbers in parentheses below give the approximate intensities, in units of 10^3 rayleighs, for the Venus observations at 13 000 km.*

Probable emitting species	Channel (Å)	Count rate (s^{-1})		
		Earth 282 000 km	Venus 194 600 km	Venus 13 000 km
Zero order	1150–1700	880	15,800	26,200 (4,000)
Zero order	200–1500	640	8,480	12,160
He	304	5	22	100
Background	430	4	15	67
He	584	100	187	233 (0.61)
Ne	740	7	21	87
A	867	17	31	100
A	1048	21	39	147
H	1216	350	506	693 (19)
O	1304	120	127	267 (17)
CO, fourth positive	1480	4	173	987 (55)
C	1657	~1	53	260 (30)

corresponding to a maximum resolution on the planet from periapsis of about 30 km. The infrared spectrometer was a circular filter wedge device, recording a 1.5-3.0 μm spectrum with a resolution of 0.1 μm in approximately 10 s. Its principal objective was to study the variations in equivalent width of the several weak CO_2 bands which fell within its range, as a function of phase angle and location on the planet. Such behaviour has been observed from the ground and is related to fluctuations in the vertical distribution of cloud opacity. Preliminary results showed a 'fuzzy' upper boundary to the cloud layer with a scale height of 3-5 km (Gnedykh *et al.* 1976).

The Lyman-α photometer used cells containing hydrogen and deuterium activated by heating a tungsten filament, which converts the molecules into atoms and so renders them opaque to Lyman-α radiation. A preliminary interpretation of the data by Kurt *et al.* (1976) shows that hydrogen and not deuterium is primarily responsible for the emission; the ratio D/H is ∼ 5 x 10^{-3}. The results are also, according to the same authors, in poor accord with existing models of

Fig. 14.8 Block diagram of the Mariner 10 DSS 14 *S*- and *X*-band radio system.

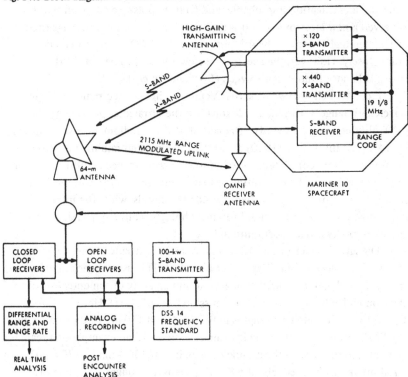

Table 14.3. *Venera 9 and 10 orbiter remote sensing instruments*

Instrument	Spectral Range
Panoramic TV Camera	0.4 μm
Spectrometer	0.24 – 0.7 μm
IR Spectrometer	1.7 – 2.8 μm
IR Radiometer	8 – 28 μm
UV Photometer	0.35 μm
Photopolarimeter	0.4 – 0.8 μm
Radio Occultation	8 cm and 30 cm

the thermal structure of the Venusian exosphere, suggesting that previously unconsidered processes may be at work.

The UV/visible spectrometer was a grating device with a spectral resolution of approximately 20 Å (0.02 μm). Fig. 14.9 shows the ultraviolet spectrum of the night side of Venus, showing the band of a species in the exosphere. Kransnopolsky *et al.* (1976) shows that, of the species for which laboratory spectra exist, the features most closely resemble BeO. The fit is not exact, however, and the identification improbable, and Kransnopolsky concluded that he had observed a new electronic transition of CO_2. Later Slysh (1976) suggested that calculated high-J lines in the 4th positive band of CO fit the observed spectrum closely. However, the spectrum is now thought to be that of the Hertzberg II bands of molecular oxygen (Lawrence *et al.* 1977).

Fig. 14.10 shows swaths across Venus by the infrared radiometer (in the wavelength range 3–13 plus 18–28 μm) and the ultraviolet photometer (at 3500 Å), obtained simultaneously. The layout of the ir radiometer itself appears in Fig. 14.11. Ksanfomality (1980) notes that maxima and minima in the two swaths are sometimes correlated, although the correspondence is subtle. Current thinking is that the ultraviolet contrast is mainly due to variations in the abundance of sulphur dioxide at some depth inside the clouds, while the infrared scans are most sensitive to the cloud-top morphology. Thus any physical relationship between the two is at most quite indirect.

The later Venera 11 and 12 missions carried extreme ultraviolet spectrometers of advanced design (Bertaux *et al.* 1981). This used a mechanical collimator, an holographic diffraction grating and an array of ten detectors of the electron multiplier type located at emission lines of H, He, He$^+$, O, O$^+$, Ar, Ne, C, and CO at discrete wavelengths between 30 and 170 nm (0.03 to 0.17 μm). Fig. 14.12 shows the layout of the instrument and 14.13 an example of a scan across Venus, in this case in the atomic oxygen line at 130.4 nm (0.1304 μm). The total intensity observed, about 6400 Rayleighs, is thought to be due primarily to resonant scattering of solar photons by oxygen atoms present in amount of a few percent at altitudes around 140 km. Other contributions to the emission, of uncertain magnitude, are excitation of O by energetic electrons and production of excited O by photodissociation of CO.

The preceding discussion shows that the atmosphere of Venus is presently of great interest not only because of its superficial similarity to our Earth, but also because it is so rich in interesting phenomena, many of which are unexplained. Outstanding amongst these are the dynamics of the upper atmosphere in general and the origins of the 4-day circulation in particular, the structure and chemistry of the clouds, and the mechanisms responsible for the high temperatures and pressures in the deep atmosphere. The recent mission to Venus by the USA was a multi-pronged assault on these and other problems by means of an orbiter and four entry probes. Of the orbiter payload, for present purposes the experiments of interest are the cloud photopolarimeter, the ultraviolet spectrometer, the infrared radiometer, and the radio occultation exper-

Fig. 14.9 (Top) Ultraviolet spectrum of the night side of Venus showing the airglow spectrum (Krasnopolsky *et al.*, 1976). The wave-length scale is in Angstroms (10^4 Å = 1 μm) and the broken line represents the instrument sensitivity vs. wavelength. The spectrum below is the Hertzberg II band of molecular oxygen (Lawrence *et al.*, 1977).

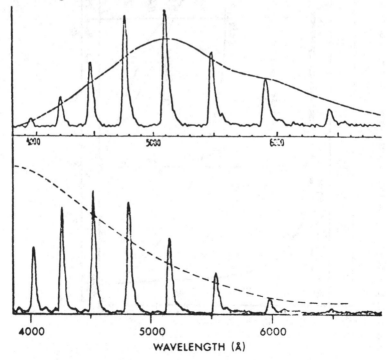

Fig. 14.10 Coincident ultraviolet and infrared scans across Venus, obtained by Venera 9 on 28 October 1975. (from Ksanfomality, 1980).

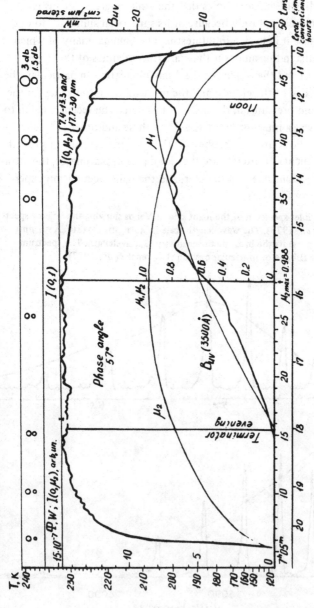

Fig. 14.11 Optical schematic of the Venera 9 and 10 radiometers, showing the arrangement for chopping incoming radiation against cold space. The chopper wheel M is driven by a motor EM and alternately allows radiation for the two sides in the mirror OP to reach the detector B. The device at the bottom of the chopper wheel is a photocell and light source used to generate a reference signal for the phase-sensitive amplifier UV (from Ksanfomality, 1980).

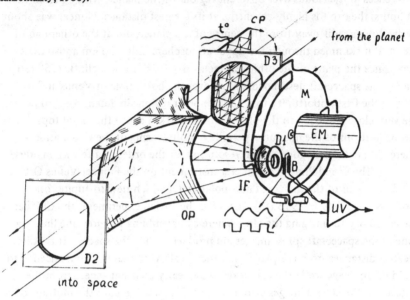

Fig. 14.12 Optical schematic of the extreme ultraviolet spectrophotometer flown on Veneras 11 and 12 (Bertaux *et al.* 1981). The numbers on the slits are the wavelengths in manometers of the energy focussed there; a separate detector lies behind each slit.

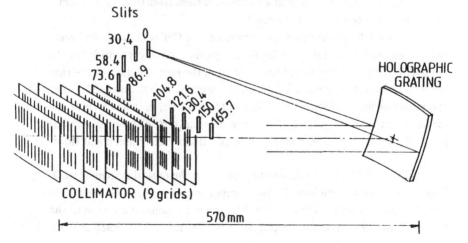

iment. The three optical instruments were mounted together on the instrument platform of the spacecraft, looking out in the same radial direction from the spin axis. This makes the data easier to correlate with each other, in the search for temperature contrasts associated with the ultraviolet markings and structural differences in the clouds over different regions of the planet. With a period of 24 hours, the orbit is highly eccentric; at its greatest distance, Pioneer was about twelve Venus radii away from the centre of the planet, but at the other end of the orbit it skimmed through the upper atmosphere only 150 km above the surface. Since the plane of the orbit was inclined at 105° to the ecliptic (75° retrograde), the spacecraft descended almost over the north pole of Venus and so offered the first opportunity to study the polar regions in detail. Also, of course, the very close approach to the planet – only 100 km above the cloud tops during portions of the mission – permits remote sensing of the upper atmosphere with extremely good spatial resolution, of the order of a few kilometers.

The Cloud Photo-Polarimeter instrument on the Pioneer Venus Orbiter is a descendant of the Imaging Photopolarimeter which flew to Jupiter on Pioneers 10 and 11 in 1973 (Ch. 16). Like the Jovian experiment, the CPP has the capability for imaging the atmosphere by assembling pictures one line at a time as the spacecraft spins. Images are produced while the spacecraft is at its greatest distances from the planet, in order to allow the several hours needed to build up an image while the planet retains a nearly constant aspect from the spacecraft. Successive images or portions of images will record the motions of clouds in a similar manner to the Mariner 10 TV experiment. The disadvantage of long 'exposure time' is offset by the advantage of several Venus years of coverage. For imaging only, a very narrow field of view is employed (0.23° × 0.29°), giving a spatial resolution of about 15 km on the planet when viewing from apoapsis. Imaging is done in a wavelength band from 0.30 to 0.39 μm, where the cloud contrasts are strongest.

During its passage close by the planet, the CPP performs a different type of remote sensing function. By measuring not only the intensity but also the polarization of the reflected radiation from the planet, details of the cloud microstructure can be obtained. 'Microstructure' refers to the mean size, shape and refractive index of individual cloud and haze particles. It was by measurements similar to those of the CPP, but made by Earth-based observers, that the gross properties of the Venus clouds were first discovered (Hansen and Hovenier 1974).

Fig. 14.14 shows a schematic diagram of the optical system of the CPP. From this, it can be seen how the two components of polarization are obtained using a Wollaston prism. The rotating filter wheel contains, in addition to the 0.30-0.39 μm filter used for imaging, 0.255-0.285 μm, 0.350-0.380 μm. 0.545-

0.555 μm and 0.945-0.955 μm filters for photopolarimetry. The detectors are two UV-enhanced silicon photodiodes operated in the photo-voltaic mode. Inflight calibration is accomplished using a solar diffuser and a standard lamp, the latter visible in the diagram.

The results from CPP imaging (Rossow *et al.* 1980) have confirmed some Mariner 10 findings but are at variance with others, possibly due to long-term changes in the atmosphere. For example, ultraviolet features of similar morphology to those in the Mariner 10 images are seen rotating with similarly rapid zonal velocities, but the equator-to-pole variations in this velocity are different. Furthermore, there are significant differences between the two hemispheres in the data from the later mission. Since the rotation axis of Venus is nearly perpendicular to the ecliptic plane, and its orbit nearly circular, these asymmetries plus the time-dependent changes suggest to Rossow *et al.* (1980) a state of 'vascillating equilibrium', with a time scale of the order of 100-1000 days, rather than a steady state, for the momentum balance in the middle atmosphere of Venus.

Polarimetry observations by the CPP have detected a submicron haze in addition to the micron sized droplets which make up the main cloud top (Kawabata *et al.* 1980). The small particles have a mean diameter of 0.23 ±0.04 μm and occur mostly in the polar regions, where they form polar caps having substantial optical thickness in the ultraviolet portion of the spectrum. Smaller amounts of haze are present at lower latitudes and diurnal and longer time scale variability has been observed.

The Pioneer Venus Programmable Ultraviolet Spectrometer (PUVS) was a development of the UV instruments which flew to Mars on Mariners 6, 7 and 9 (Ch. 15). Its objectives were threefold: (1) to measure the intensity of Lyman-α emission from atomic hydrogen in the exosphere; (2) to measure the spectrum of the UV airglow in the thermosphere; and (3) to measure the UV spectrum of scattered sunlight from clouds and hazes in the middle atmosphere. To deal with these in reverse order, it was hoped that the cloud observations could provide the key to understanding the ultraviolet contrasts in the clouds, since UV spectra of individual light and dark regions were to be obtained. At the same time, the PUVS had the capability for detecting and measuring the distribution of ozone on Venus, if significant quantities are present. This was a crucial search, in view of the importance of ozone in the Earth's atmosphere (Ch. 10); however, none was detected.

In the thermosphere, spectroscopy of the UV airglow reveals the vertical and global distributions of CO_2, CO_2^+, CO, O, H and C, and searches for the presence of N_2 or N. A crucial question here is that of the stability of CO_2 on Venus. The excitation mechanisms for these emissions and the temperatures in

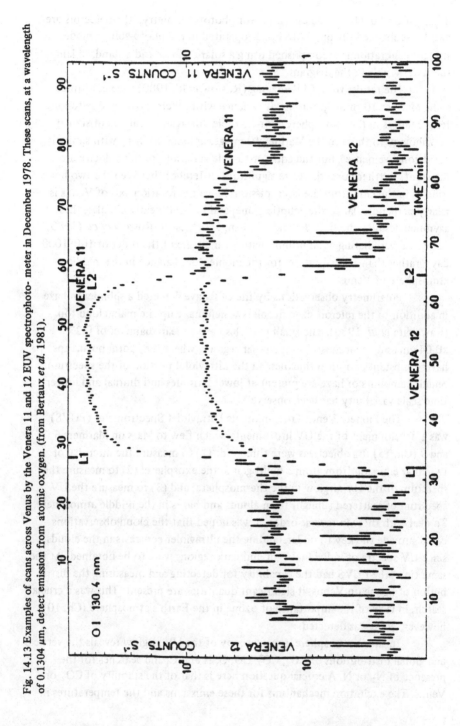

Fig. 14.13 Examples of scans across Venus by the Venera 11 and 12 EUV spectrophoto- meter in December 1978. These scans, at a wavelength of 0.1304 μm, detect emission from atomic oxygen. (from Bertaux et al. 1981).

Fig. 14.14 Pioneer Venus Orbiter photopolarimeter optical schematic (from Colin and   Hunten, 1977).

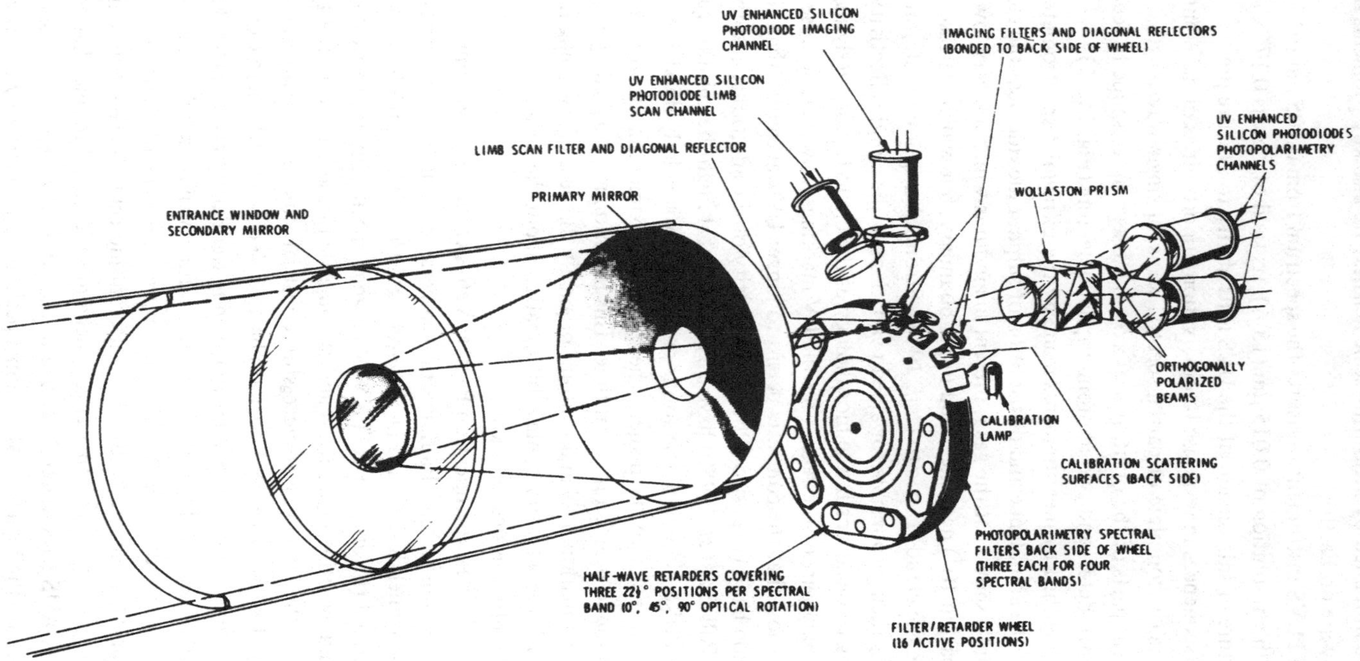

UV ENHANCED SILICON PHOTODIODES PHOTOPOLARIMETRY PHOTOPOLARIMETRY CHANNELS

IMAGING FILTERS AND DIAGONAL REFLECTORS (BONDED TO BACK SIDE OF WHEEL)

ORTHOGONALLY POLARIZED BEAMS

WOLLASTON PRISM

CALIBRATION LAMP

CALIBRATION SCATTERING SURFACES (BACK SIDE)

PHOTOPOLARIMETRY SPECTRAL FILTERS BACK SIDE OF WHEEL (THREE EACH FOR FOUR SPECTRAL BANDS)

FILTER/RETARDER WHEEL (16 ACTIVE POSITIONS)

UV ENHANCED SILICON PHOTODIODE IMAGING CHANNEL

UV ENHANCED SILICON PHOTODIODE LIMB SCAN CHANNEL

LIMB SCAN FILTER AND DIAGONAL REFLECTOR

PRIMARY MIRROR

HALF-WAVE RETARDERS COVERING THREE 22½° POSITIONS PER SPECTRAL BAND 10°, 45°, 90° OPTICAL ROTATION)

ENTRANCE WINDOW AND SECONDARY MIRROR

the exosphere (obtained by measuring scale heights) are also key problems on Venus as they are on Earth.

The PUVS instrument measures the spectrum from 0.105 μm to 0.340 μm with a resolution of 0.015 μm (15 Å). Its field of view is 0.17° x 1°. A special feature of this version of the UVS is its ability to execute pre-programmed sequences in response to observation opportunities of different kinds, i.e. to select complete or partial spectral scans, or single wavelengths, during different parts of the orbit. It uses a 5 cm Cassegrain telescope to feed an Ebert spectrometer, and two photomultiplier detectors (Fig. 14.15).

Among the key measurements made by the ultraviolet spectrometer were maps of the abundance and distribution of sulphur dioxide near the cloud tops. The results show that the gas is present in amounts which are very low (about one part in 10^7) near the cloud upper boundary, but which increase rapidly inside the cloud (Esposito *et al.* 1979). Strong absorption by SO_2 at UV wavelengths is then responsible for albedo variations in the cloud reflectivity corresponding to variations in the morphology of the cloud top or overlying haze. This is only part of the reason for the UV markings, however, because the observations show that the contrasts persist to longer UV wavelenths than the SO_2 absorption band (Esposito 1980). The identity of the additional absorber is unknown. Pollack *et al.* (1980) have suggested that it might be chlorine gas.

The PUVS airglow observations produced a particularly important result on the night side of the planet, where a local maximum in the nitric oxide emission was found near 2 a.m. local time. This radiation is produced by the radiative recombination of N and O atoms originally dissociated by sunlight, so maps of its darkside intensity contain information about the circulation of the atmosphere at altitudes near 115 km.

The infrared experiment on Pioneer Venus carries the acronym VORTEX, for Venus Orbiter Radiometric Temperature-sounding Experiment. It is an eight-channel device, ranging in wavelength from the ultraviolet (0.4 μm) to the far infrared (55 μm). VORTEX was designed specifically for vertical temperature sounding, utilizing narrow-band radiometric measurements in the 15 μm band of CO_2 in the same way as successful terrestrial atmosphere temperature sounders (Fig. 14.16). Because the atmosphere of Venus is nearly pure CO_2 the vertical coverage obtained is much greater than in the Earth's atmosphere. Fig. 14.17 shows the computed weighting functions, for a Venus model atmosphere based on Mariner 5 radio occultation temperature measurements. The three uppermost curves are generated by three different molecular sieve settings in a single pressure modulator radiometer channel almost identical to one channel of the Nimbus 7 SAMS radiometer described in Ch. 10. In the Venus atmosphere, viewing vertically, the highest PMR channel peaks at the remarkably low pressure

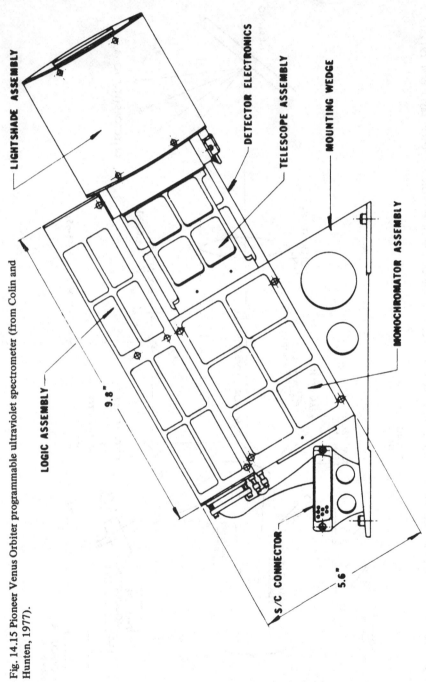

Fig. 14.15 Pioneer Venus Orbiter programmable ultraviolet spectrometer (from Colin and Hunten, 1977).

262

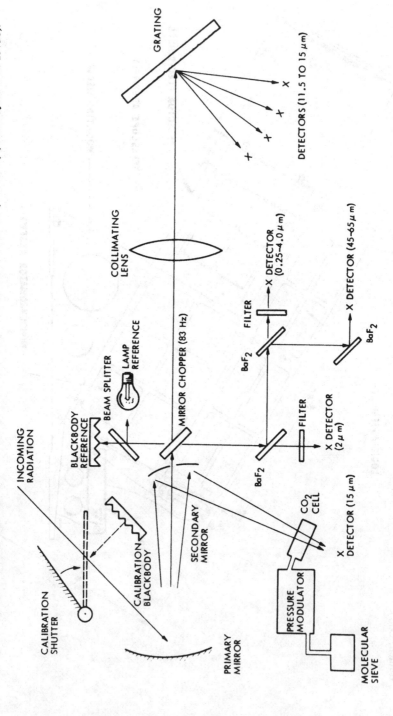

Fig. 14.16 Schematic diagram of the optical system of the Pioneer Venus Orbiter infrared radiometer (VORTEX) (from Taylor *et al.* 1979b).

Fig. 14.17 Weighting functions for the seven VORTEX temperature sounding channels. The origin of the radiance measured in each channel is distributed vertically in proportion to the value of the weighting function at each level. This effect determines the vertical resolution of the temperature profile retrievals. One example of the effect of Doppler shift on the channel 1 (PMR) weighting functions is shown (from Taylor *et al.* 1980).

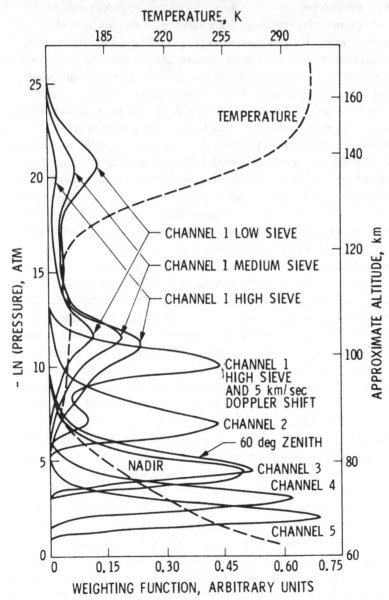

of 10^{-9} atmospheres, well inside the ionosphere and actually just a few kilometers below the lowest pass of the orbiter spacecraft. Previous temperature measurements on Venus have always been made in atmospheric 'windows', corresponding to maps of the cloud-top temperature. These showed intriguing contrasts, waves and 'hot spots', but were difficult to interpret because of the problem of distinguishing real temperature variations from changes in cloud height or distribution. The principle followed by VORTEX is to measure temperature distribution on seven constant-pressure surfaces, revealing the thermal contrasts at each level which drive the complex motions seen in the ultraviolet images, and allowing individual profiles to be retrieved (Fig. 14.18). Like the CPP, VORTEX used very fast integration times (0.024 s in this case) to build up complete pictures of the atmosphere from horizon to horizon as the spacecraft spins. The first significant finding from these images came when the polar region was viewed for the first time (Taylor *et al.* 1979a). Then it was seen that the uniform cloudiness on Venus is broken near the pole, where the atmosphere descends rapidly in a giant vortex. More surprisingly, the 'eye' of the vortex has a remarkable double appearance, and is surrounded by a crescent-shaped collar of cold atmosphere. These phenomena are planetary-scale wavelike perturbations in the general circulation, probably instabilities in the zonal flow as it moves polewards with a consequent tendency for its angular momentum to increase.

The problems of temperature-sounding in the presence of cloud, discussed in Ch. 8 for the case of the Earth's atmosphere, is particularly complex for Venus where there is total cloud cover and little *a priori* knowledge of the vertical distribution of clouds. Fortunately, the complex geometry of the Pioneer Venus orbit results in views of the planet over a complete range of zenith angles (angle between the line of sight of the instruments and the local vertical at the viewed point on Venus). The variation of the CO_2 opacity with zenith angle is amenable to calculation, allowing cloud opacity contributions to be separated (Taylor 1974; Schofield and Taylor, 1983). At zenith angles close to 90°, this technique becomes the same as limb scanning (Ch. 6). Like the PUVS and CCP, VORTEX has a sufficiently narrow field of view (1.25°) that several resolution elements can be obtained viewing through the atmosphere tangentially. Internal logic reduces the dwell time of the radiometer to only 0.012 s during a limb crossing, resulting in points separated by about 2 km in the vertical direction. From these, additional information on the vertical distribution of temperature and clouds was obtained when the spacecraft was near periapsis. Among the key findings from temperature measurements was the fact that the polar atmosphere is warmer than that near the equator over the altitude range from about 70 to 90 km (Taylor *et al.* 1980; Fig. 14.19). This produces pressure forces which tend to decelerate the zonal flow. This means that the $100\,\mathrm{km s^{-1}}$ winds observed in

Fig. 14.18 Vertical profiles of temperature versus pressure at 30°N latitude as measured on Venus by VORTEX (solid line) and Earth by Nimbus 7 Stratospheric and Mesospheric Sounder (dashed line). From Taylor *et al.* (1981).

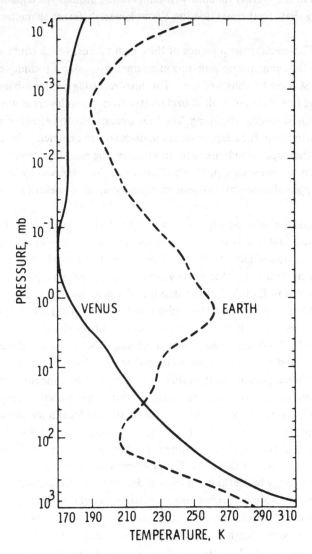

the motions of the UV markings near the cloud tops probably reduce sharply in speed at higher levels (Elson 1979). Another remarkable finding was the existence of two maxima in the diurnal variability of temperature around the equator (Fig. 14.20); showing that the thermal tides on Venus have a different character to those on Earth.

VORTEX detects the presence of thin, high clouds, which could affect the integrity of the temperature sounding measurements, using a technique first applied by the SCR on Nimbus 5 (Ch. 6). This involves using a narrow-band near-infrared channel ($\lambda = 2\,\mu m$) to look at backscatter from cloud layers at a wavelength where CO_2 is weakly absorbing. The CO_2 opacity has the effect of masking the backscatter from the deep, main cloud deck and so enhancing the contrast by which the upper clouds are seen. In practice, the only high cloud found was an extremely tenuous haze, probably the same as that observed by the CPP. Its opacity at infrared wavelenths is vanishingly small, however (Taylor *et al.* 1980).

The question of water vapour on Venus has long been a puzzle. The amounts observed from Earth are sometimes too large to be in equilibrium with the sulphuric acid clouds, and the distribution appears to be patchy and time dependent. The implication is that water vapour is a key component of Venus' meteorology, as it is in Earth's, but the absence of oceans and presence of strongly hygroscopic clouds make the cycles which it follows quite different. As a first attempt to establish the patterns and behaviour of water vapour on Venus, VORTEX carried a 35–45 μm channel which measures the opacity of the atmosphere in the strongest part of the pure rotational band of H_2O. It found only very small abundances, much less than some of the Earth-based measurements and orders of magnitude less than at the same pressures in our own atmosphere. The abundance increases by more than a factor of 10 near local noon at low latitudes, where the solar flux is greatest (Fig. 14.21; Schofield and Taylor 1982b). Probably the heat of the sun is evaporating the vapour out of the cloud droplets.

Other objectives of the VORTEX experiment were measurements of the atmospheric energy budget and the angular distribution of radiation scattered from the clouds. The former involves using a broad spectral channel (0.4–4.0 μm) to measure the backscattered solar intensity and hence determine the local albedo at the same location on the planet where temperature structure is simultaneously measured. The difference between the net ingoing solar and net outgoing thermal components determine the heating, which is greatest near local noon at the equator. This energy is redistributed by the atmospheric circulation and reradiated to space at all locations. The atmosphere is thus an enormous heat engine, and the balance between radiant and dynamically transported energy at each latitude and time of day is a crucial puzzle. The results (Tomasko *et al.*

Fig. 14.19 Zonally averaged mean meridional temperature in the middle atmosphere of Venus, retrieved from VORTEX data (Schofield and Taylor, 1983).

RETRIEVED ZONAL-MEAN VENUSIAN TEMPERATURE FIELD.

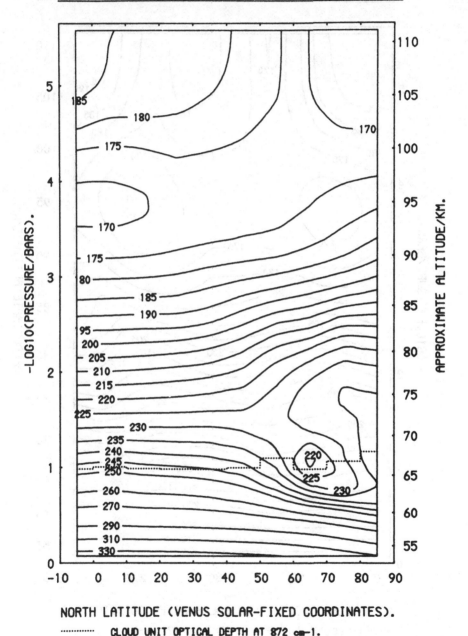

NORTH LATITUDE (VENUS SOLAR-FIXED COORDINATES).

·············· CLOUD UNIT OPTICAL DEPTH AT 872 cm-1.

Fig. 14.20 Mean zonal temperatures in the middle atmosphere of Venus, retrieved from VORTEX data (from Schofield and Taylor, 1983).

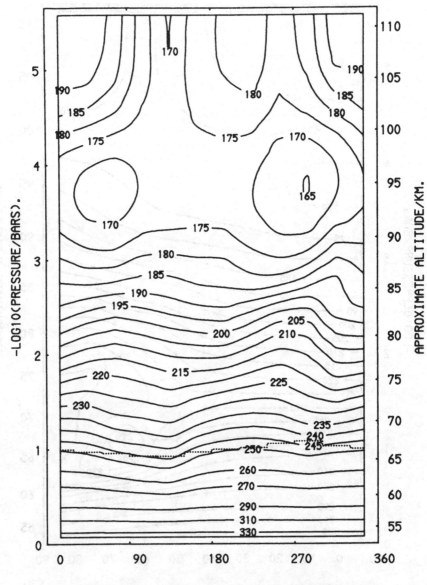

RETRIEVED MEAN VENUSIAN TEMPERATURE FIELD (0-30N).

SOLAR LONGITUDE (VENUS SOLAR-FIXED COORDINATES).

.... CLOUD UNIT OPTICAL DEPTH AT 872 cm-1.

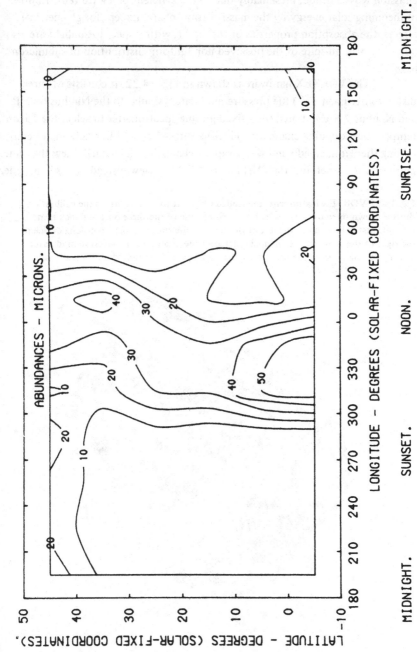

Fig. 14.21 Mean column abundances for water vapour above the cloud tops on Venus, derived from VORTEX data (Schofield and Taylor, 1982b).

1980; Schofield and Taylor, 1982a) show a very uniform profile of thermal emission with latitude, presumably due to the efficient poleward redistribution of incoming solar energy by the massive atmosphere. Except for geometrical effects, the absorption properties of the clouds with respect to sunlight are also quite uniform, in spite of the observed non-uniform distribution of submicron haze.

The VORTEX hardware is shown in Fig. 14.22. It consists of three distinct instruments: (i) the pressure modulator, similar to the Nimbus 6 PMR and Nimbus 7 SAMS units; (ii) a fixed grating spectrometer to select the 15 μm temperature sounding channels; and (iii) a three-channel filter radiometer containing the 2 μm, albedo and water-vapour channels. All channels view the same point on the planet, but the PMR has a 5° field of view instead of 1.25° in order

Fig. 14.22 VORTEX instrument: the ribbed object at the top centre is the calibration shutter shown open over the entrance aperture. One of the three optical bench heaters is visible just below and to the left of the shutter; the small, elongated oblong device to the right of the heater is the channel 8 detector preamplifier. The pressure modulator and the molecular sieve are seen below the optical bench, mounted on the baseplate (from Taylor *et al.* 1979).

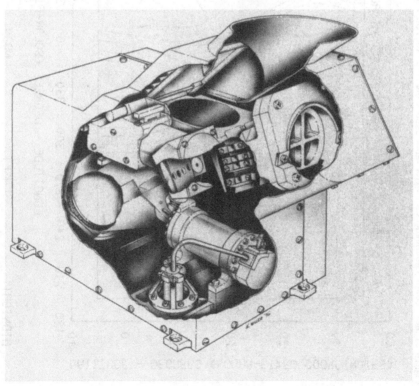

to increase its energy grasp to compensate for the extremely high effective spectral resolution. All of the detectors are deuterated tryglycine sulphate pyro-electric devices, ranging in size from 0.6 to 2.5 mm diameter. In-flight calibration is obtained by viewing cold space every 12 seconds, as the spacecraft rolls, and by occasionally closing a mirror shutter over the entrance aperture to view an internal blackbody (for the thermal channels) and a standard lamp (for the near-infrared channels). Laboratory measurements of the weighting functions were carried out using a White cell in the manner described for Nimbus temperature sounders in Ch. 6. A major difference, however, is the need to obtain a much larger path length of CO_2 to simulate the Venus atmosphere – 1000 m atm are needed compared to 0.25 for Earth. Thus, the White cell had to be extremely large, and coolable to approximately 220 K.

     The Pioneer Venus radio occultations used S and X bands to obtain two profiles during each 24 hour orbit (except, of course, for those seasons when the orbiter does not pass behind the planet as seen from Earth). Thus, broad geographical coverage was obtained during the course of the mission. The occultation results were the first to identify strong inversions in the high-latitude temperature profiles (Fig. 14.23; Kliore and Patel 1980). In general, they agree very well with the profiles derived from infrared remote sounding.

Fig. 14.23 Temperature profiles on Venus obtained by the radio occultation method using the PV orbiter (Kliore and Patel, 1980).

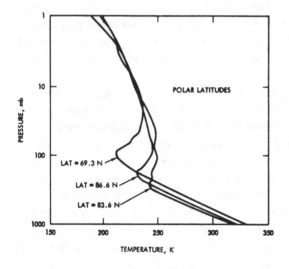

15

REMOTE SOUNDING OF THE ATMOSPHERE OF MARS

Mars, the first of the superior planets, has an atmosphere about one hundred times less thick than that of the Earth and ten thousand times less thick than that of Venus. The Martian atmosphere is almost free of clouds where the terrestrial atmosphere is about 50% cloudy and Venus nearly 100% covered. The surface topography of Mars is more extreme than that of Earth, and so influences the atmospheric circulation more; the Venus circulation in the observable atmosphere is probably virtually free of the influence of surface topography because of the massive blanket of deep atmosphere between the two. Finally, Mars has no seas of liquid water; Earth is 85% covered and Venus is effectively 100% covered since the lower atmosphere is like an ocean insofar as it behaves as a nearly infinite source or sink of heat and momentum for the upper atmosphere. Thus Mars is the third of a set of terrestrial planets with graduated properties with the Earth in the middle in each case. Naturally the Martian atmosphere is a topic of intense interest and has already been the target of numerous remote sensing investigations.

Mariner 4 was the first successful Mars probe, arriving in mid-1965. Television cameras were carried, but no infrared or ultraviolet instruments, and, other than detection of thin clouds, the atmospheric investigations were mostly restricted to an S-band radio occultation experiment, similar to the Mariner 5 measurements at Venus (Ch. 14). The precision with which the raw data must be corrected for effects due to the Earth's atmosphere and to the relative motion of the probe, planet and receiving station is particularly demanding for the signals from Mars since the atmosphere is relatively tenuous. For the Mariner 4 experiment Kliore (1968) estimates that the contribution of all sources other than the Martian atmosphere were known to better than 1 part in 10^{11}. The results indicated surface pressures of 5 mbar on occultation and 9 mbar on reappearance, considerably lower than had been expected from Earth-based spectroscopy.

Mariners 6 and 7 encountered Mars within a few days of each other in July–August 1969. These were distinguished from Mariner 4 by enhanced payloads which included infrared and ultraviolet spectrometers for atmospheric investigations. The S-band occultation experiment also yielded improved results, including extra data on the Martian ionosphere. The 1969 Mariners used the two-way ('closed-loop') mode of operation whereby a frequency referenced to a rubidium standard is transmitted to the spacecraft and coherently retransmitted. The contributions to the phase shift of the Martian neutral atmosphere and ionosphere are

$$\Delta\phi\,(\rho,\nu) = \frac{1}{\lambda}\int_{-\infty}^{\infty}\,(n-1)\,dx + \frac{1}{\lambda}\int_{-\infty}^{\infty}\,(n_i-1)\,dz \qquad (15.1)$$

where λ is the wavelength and n, n_i the refractive indices of the neutral and ionized atmosphere respectively; n_i depends strongly on frequency, according to the formula

$$n_i = 1 - \frac{4.03 \times 10^{-7}\,N_e}{f^2} \qquad (15.2)$$

where N_e (cm^{-3}) is the number density of electrons and f is the frequency in Hz, which must be very much greater than the cyclotron frequency for the validity of (15.2); while n is almost constant. This allows the two contributions to be separated by making measurements at two wavelengths, although actually they are also generally well separated in altitude. Fig. 15.1 shows Martian day-side density, temperature and ionosphere electron density profiles as obtained by Mariner 6.

The infrared spectrometer on the 1969 Mars flybys (Herr *et al.* 1972) was a variable-wedge interference filter device with approximately 1% spectral resolution and a field of view of 2°, corresponding to 120 km of spatial resolution on the planet at closest approach. The wavelength range 1.9–14.3 μm was covered, using a radiation-cooled lead selenide detector for the shorter wavelengths and a mercury-doped germanium bolometer cooled to 20 K by a Joule–Kelvin device for the longer wavelengths; wavelength calibration was achieved by taking the spectrum of an on-board polystyrene film at intervals. Fig. 15.2 shows a spectrum of Mars obtained with this device; three atmospheric constituents (CO_2, CO and H_2O) are unambiguously detected and the detailed analysis of the aggregated data was used to set upper limits on a large number of possible minor constituents (Horn *et al.* 1972).

The Mariner 6 and 7 ultraviolet spectrometer (Pearce *et al.* 1971) was designed to observe the upper atmosphere of Mars at altitudes above 130 km by viewing the limb of the planet in the spectral range 0.11 to 0.43 μm, with a spectral resolution of $0.002 \mu m$ (20 Å). The instrument (Fig. 15.3) is an Ebert scanning spectrometer, with a plane diffraction grating and photomultiplier tube detectors. A cam system drives the grating back and forwards to produce a spectrum (Fig. 15.4) every three seconds. The features in the spectrum reveal the composition of the upper Martian atmosphere and their intensities provide measurements of scale height for each species. In addition to ionized CO_2 and CO features, atomic oxygen lines were observed and an extensive hydrogen coma was discovered (Barth *et al.* 1971).

On November 14th, 1971, Mariner 9 became the first Mars orbiter, closely followed by the Russian Mars-2 and Mars-3 orbiters on November 27 and December 2 respectively. Fig. 15.5 shows the spectral ranges covered by the remote sensing instruments on Mariner 9.

One of the immediately obvious advantages of the orbiter *versus* the flyby spacecraft is the large number of radio occultations by the atmosphere. With such a large number of profiles, it becomes possible to study the atmosphere on a global scale and to observe regional and seasonal cycles in temperature and pressure (Kliore *et al.* 1972, Woiceshyn 1974).

Fig. 15.1 Electron density profiles in the ionosphere of Mars, with a terrestrial profile for comparison (from Gringauz and Breus 1970).

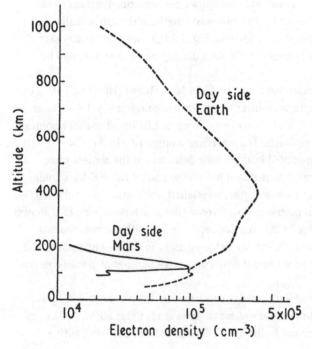

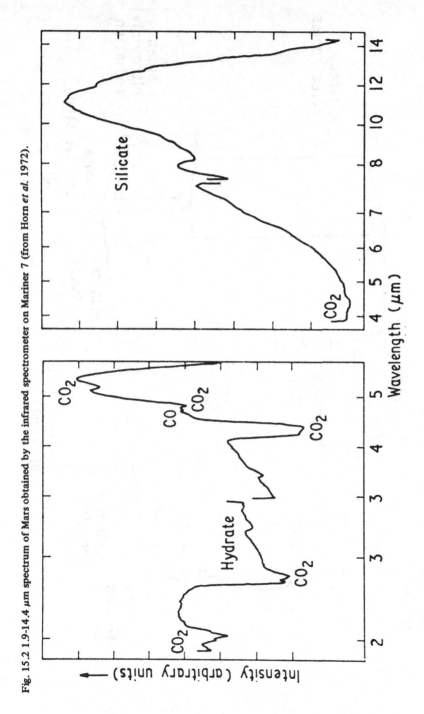

Fig. 15.2 1.9-14.4 μm spectrum of Mars obtained by the infrared spectrometer on Mariner 7 (from Horn *et al.* 1972).

Fig. 15.3 Optical layout of the ultraviolet spectrometer used on Mariners 6 and 7 (from Barth *et al.* 1969).

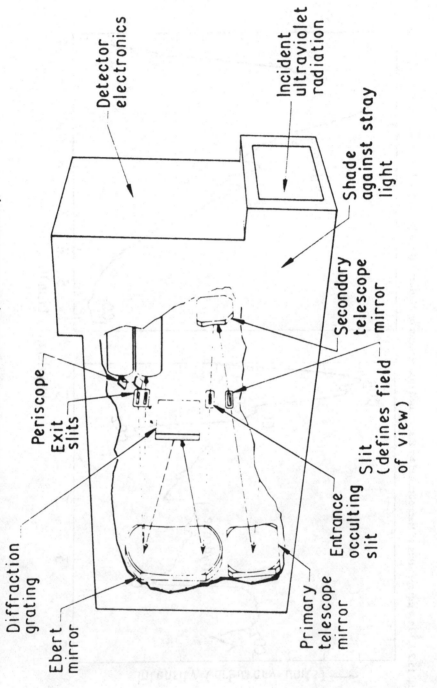

The infrared spectrometer on Mariner 9 was a Michelson interferometer, a development of the IRIS experiments on Nimbus 3 and 4 (Ch. 6). The Mariner IRIS covered the spectral range from 5 to 50 μm with a spectral resolution of 2.4 cm^{-1} (about 0.06 μm at 5 μm and 0.6 μm at 50 μm), and an angular field of view of 4.5°, corresponding to a spatial resolution of 125 km when viewing vertically from periapsis. Like the Earth-orbiting IRIS, the Mariner instrument features a thermistor bolometer detector, image-motion compensation, and in-flight calibration by means of a rotating object-space scan mirror which views cold space or an on-board black-body. Wavelength calibration was obtained by referencing all infrared spectra to the 0.6929 μm line of a neon discharge tube (Hanel *et al.* 1972). Mariner 9 IRIS spectra of Mars show the 15 μm CO_2 band in emission over the Martian poles, indicating a stratospheric temperature which is higher than that of the surface. Fig. 15.6 shows the actual temperature field

Fig. 15.4 Ultraviolet spectrum of the Martian limb taken from Mariner 7 (from Barth *et al.* 1969).

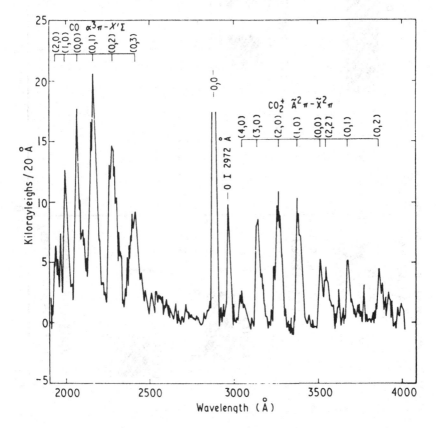

278

Fig. 15.5 Spectral coverage of the instruments on Mariner 9.

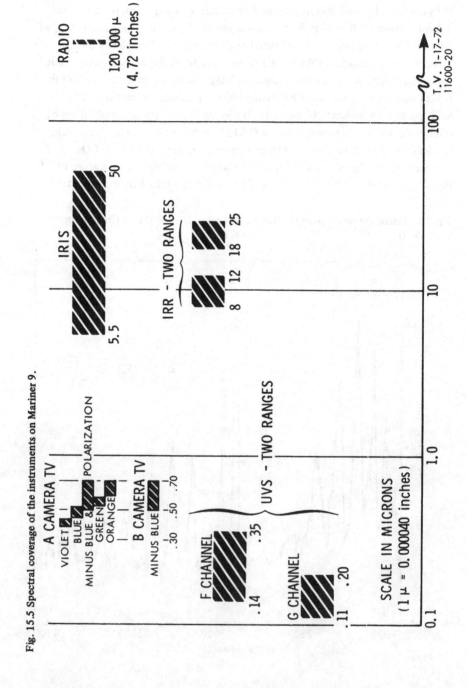

obtained by inverting 15 μm band radiances; in Fig. 15.7 appears a set of weighting functions for an instrument using filters and pressure modulators to observe a pure CO_2 Martian model atmosphere. It is interesting to compare these with equivalent calculations (but for instruments of different spectral resolutions) for Earth and Venus, shown in Figs. 6.2 and 14.17 respectively. By virtue of its extensive coverage of the planet, global temperature data has allowed preliminary studies of the general circulation on Mars (Conrath 1976, 1981). IRIS data is also rich in information on the abundance of water vapour and the composition of surface and windborne dust on Mars (Hanel *et al.* 1972; Conrath *et al.* 1973).

The Mariner 9 ultraviolet spectrometer (Hord *et al.* 1970) was similar to the Mariner 6 and 7 devices described above. In addition to observations of CO,

Fig. 15.6 Mean meridional temperature cross-section for the N hemisphere of Mars, retrieved from Mariner 9 IRIS data (Conrath, 1981).

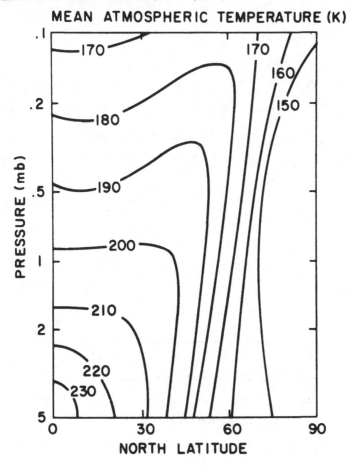

Fig. 15.7 Temperature sounding weighting functions for the Martian atmosphere (J.T. Schofield, personal communication).

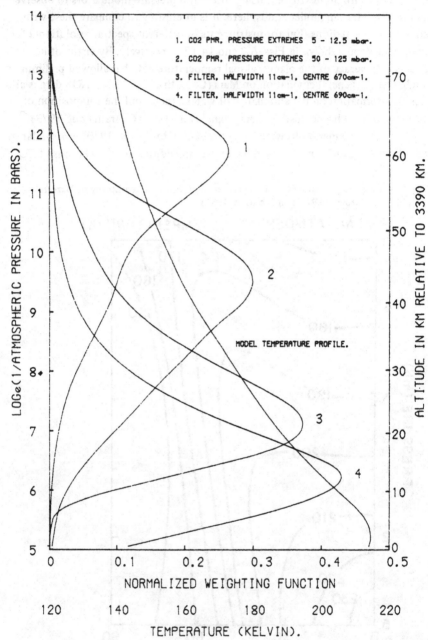

O and H in the airglow (Barth *et al.* 1972), this instrument mapped the surface pressure on Mars by measuring the local variations in Rayleigh scattering intensity, which is a measure of the total molecular density. Of particular interest are the UVS observations of Martian ozone (Lane *et al.* 1973). Mariner 7 had earlier detected ozone on the south polar cap during its late spring. The early Mariner 9 observations found ozone only near the north polar cap, until later in the mission when it appeared southward of 50° S as the Martian autumn equinox approached. Lane *et al.* (1973) concluded that the abundance of ozone is anticorrelated with that of water vapour.

The television cameras on Mariner 9 recorded a vast number of interesting atmospheric phenomena on Mars, including the great global dust storm which persisted for the whole of the first month in orbit (Leovy *et al.* 1972). Clouds, probably consisting of water-ice in some cases and frozen CO_2 in others, appear frequently as consequences of the surface morphology, for example as the Martian analogue of terrestrial lee waves. The diffuse brightening north of 45°N, known as the north polar band, was found to consist most probably of CO_2 ice clouds in the atmosphere, which on detailed examination show large day-to-day variations. Finally, a persistent thin cloud layer, probably of water-ice, was observed near the 0.02 mbar pressure level. Similar layers ('noctilucent clouds') occur in the Earth's atmosphere.

The Russian spacecraft Mars-2 and 3, which were operating in orbit around the planet at the same time as Mariner 9, carried five remote sensing instruments. One of these, an 8-40 μm infrared radiometer, was, like the Mariner 9 8-12 μm plus 18-25 μm two-channel instrument, designed for thermal mapping of the surface, as was the 3.4 cm wavelength microwave radiometer carried on board the Soviet spacecraft. A narrow band photometer observed the 2.06 μm CO_2 band for the purpose of measuring the surface pressure and hence the planetary relief (Moroz and Ksanfomality 1972). A 40 mm diameter Cassegrain objective focuses the planetary radiation onto two lead sulphide photoconductive cells. The beam is split by a rotating mirror chopper, in one arm of which is a 2.25 μm filter located in the continuum adjacent to the CO_2 band. The other arm contains a wheel with 2.017, 2,050, 2.067, 2.075 and 2.25 μm filters, each 0.02 μm wide. In this way, the profile of the CO_2 band is measured independently of the target brightness. From laboratory measured equivalent widths, the total CO_2 path in the Martian atmosphere is determined and converted to a surface pressure. The instrument's field of view is approximately 0.6°, corresponding to 15 km on the planet from closest approach.

The column abundance of water vapour was measured by observing reflected near-infrared solar radiation at wavelengths near the 1.38 μm band of water vapour, collected by an 80 mm diameter Cassegrain telescope and directed onto a PbS detector via a narrow-band ($0.06\,\mu$m) interference filter, a fixed polaroid, a birefrigerant plate made of calcite, a rotating polaroid, and filters of silicon and ammonium dihydrogen phosphate (Moroz and Ksanfomality 1972). With this arrangement, the signals from the water vapour and the continuum are obtained from the same detector with phases which are 90° apart, and hence separable by synchronous detection. The instrument is sensitive to as little as 0.1 precipitable microns of water vapour. Amounts ranging from 0 to 20 pr. μm were observed on the planet, the maximum occurring in the equatorial region.

In the 1973 launch opportunity, two further Soviet spacecraft, Mars-5 and 6, were placed in orbit about Mars. The payloads included a water vapour spectrometer similar to that on Mars-3; on this occasion (February 1974) much larger amounts of water vapour (up to 100 pr.μm) were observed, even though the Martian season was the same. A 3000–8000 Å spectrometer, with 20 Å of resolution, searched in vain for a Martian night airglow. The absence of emissions places upper limits on several ionospheric species, such as H, O and O_2^+. The infrared radiometer had the same spectral range as Mars-2 and 3 (8–40 μm, with the 13–17 μm region blocked out to remove the 15 μm CO_2 band) and so was a surface-mapping experiment. So was the 4.3 cm wavelength microwave radiometer. Finally, Mars-5 carried an improved version of the filter-wheel 2 μm CO_2 band spectrometer for measuring the intensity of the CO_2 absorption and hence the surface pressure. This used cooled silicon detectors and obtained a surface resolution of 200–300 m.

The 1976 Viking missions to Mars, as is well known, were principally conceived in order to place instrumented landers on the planet's surface. However, the orbiter portions of the two spacecraft were each equipped with colour television, an infrared radiometer and an infrared spectrometer, and the first dual-frequency S and X band radio occultation experiment for Mars. As on Mariner 9, the imaging experiment (Carr *et al.* 1972) has revealed a wealth of detail concerning atmospheric motions and condensates (Briggs *et al.* 1979).

The Mars Atmospheric Water Detector (MAWD) on the Viking orbiters is a grating spectrometer operating in the 1.4 μm bands of water vapour (Farmer and LaPorte 1972; Farmer *et al.* 1976). Radiation from a single grating is dispersed onto five radiatively cooled lead sulphide detectors, corresponding in position to the wavelengths of three strong lines near the band centre and two continuum wavelengths. The spectral resolution of the instrument is 0.0002 μm, corresponding to a detection capability of 1 pr. μm of water vapour. A capability is incorporated for scanning the grating over an angular range corresponding to about 0.002 μm to verify the positions of the absorption features and the spectral resolution of the instrument. The field of view of the MAWD is approximately 1° × 0.1°, corresponding to a spatial resolution on the planet of

3 × 24 km when viewing from periapsis. A single measurement sequence consists of a raster of 15 contiguous fields of view, obtained in 4.48 s by means of a scanning mirror external to the 2.5 cm entrance aperture of the telescope. In-flight calibration is achieved by means of tungsten lamps.

Fig. 15.8 shows a schematic diagram of the MAWD optical layout. Fig. 15.9 shows the portion of the water-vapour spectrum which is covered by the five channels. The three line channels stand at the centres of lines differing in intensity by more than a factor of ten, and hence allow the three independent parameters, water-vapour amount, mean pressure and temperature, to be determined under sufficiently favourable conditions (Farmer and LaPorte 1972). This in turn allows an estimation of the mean height of the water vapour above the surface.

Observations of Mars with the MAWD on Viking 1 have revealed an extremely dry southern hemisphere (0–3 pr. μm, Farmer *et al.* 1976). A gradual increase across the equator leads up to a maximum of 75 pr. μm near the north pole, which was enjoying local midsummer at the time (Fig. 15.10). Most importantly, the water-vapour abundances above the northern residual (summertime) polar cap require near-surface temperatures higher than 200 K in order to prevent the water from condensing out, and in fact the IRTM (see below) measured temperatures close to 205 K (Kieffer *et al.* 1976). These measurements together rule out a solid carbon dioxide composition for the cap, which must be water-ice. Thus an ancient debate is finally settled.

The Viking Infrared Thermal Mapper (IRTM) was a six-channel radiometer which measures thermal emission from the surface of Mars at four wavelengths from 6–24 μm, thermal emission from the stratosphere at 15 μm, and reflected solar radiation in the 0.3–3 μm band (Kieffer *et al.* 1972, 1977). The instrument used thermopile detectors and had an angular field of view of approximately 0.3° (about 10 km on the planet from periapsis). In addition to its role in the identification of the north polar cap composition (Kieffer *et al.* 1977) and many observations of a geological nature, the IRTM has observed the diurnal and areographic variations of the mean Martian stratospheric temperature (Kieffer *et al.* 1977). The variation in the mean temperature of the 10 to 0.1 mb layer is of the order of 10 K during a diurnal cycle at low latitudes. In contrast, the surface below varies daily by 100 K or more. This behaviour is in accordance with theoretical models. Near the winter pole on Mars, the surface temperatures observed by IRTM are so low that they fall below the condensation temperature of CO_2. As a result, the major constituent of the atmosphere condenses on to the winter polar cap and the mean surface pressure drops planet-wide by about 25%. The condensation of all of the atmospheric CO_2 is prevented by the presence of non-condensible argon and nitrogen which take over as major constituents at the

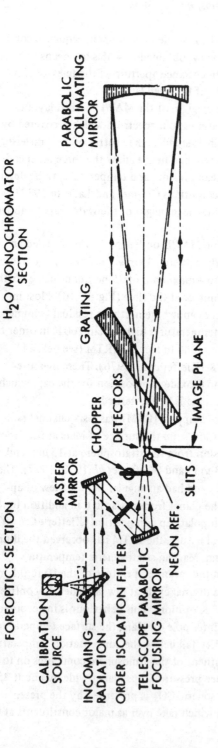

Fig. 15.8 Viking MAWD optical configuration (from Farmer *et al.* 1976).

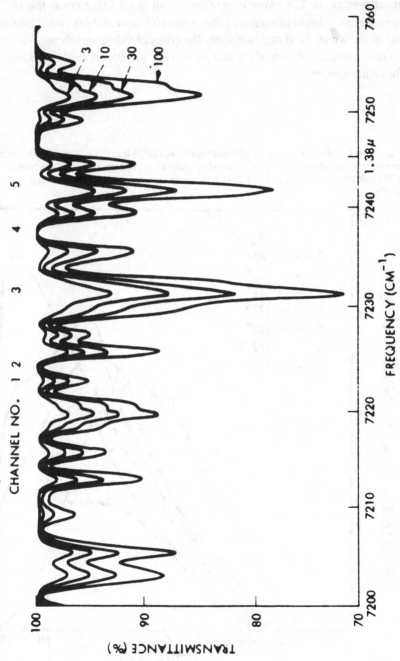

Fig. 15.9 Theoretical water-vapour spectra in the frequency range 7200-7260 cm^{-1}. The spectral positions of the five detectors are indicated by the numbers 1-5. The curves are for vapour of 3, 10, 30 and 100 μm at a total pressure on 6.9 mbar and a temperature of 225 K.

winter pole and through which CO_2 from lower latitudes must first diffuse before it can precipitate. The winter is too short for all of the CO_2 to meet this fate; nevertheless, it has been estimated that some 10^{13} tons of it are cycled into and out of the winter polar cap each year. The effect of this titanic climate variation on the general circulation of the atmosphere is a question of major significance for future missions.

Fig. 15.10 Latitude distributions of water vapour at 180° W longitude. The column abundances are averaged over 10° of latitude and are taken from observations made between 1200 and 1400 h local time. The season is northern summer (planetocentric longitude - 108°).

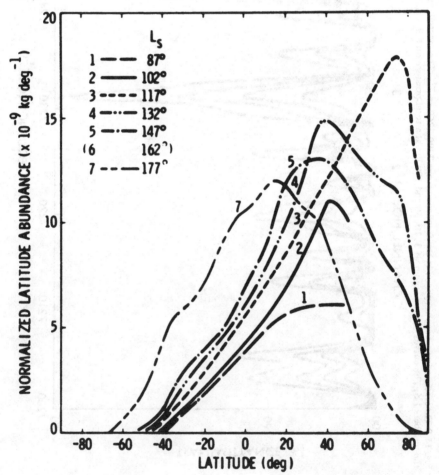

16

REMOTE SOUNDING IN THE OUTER SOLAR SYSTEM

Jupiter is more massive than all of the other planets taken together. Its vast, deep atmosphere is a fascinating whorl of multicoloured clouds and curious dynamical and meteorological phenomena. It is a natural target for investigation by atmospheric scientists using remote sensing as a tool. In the few years since missions spanning the vast distances to Jupiter became possible, we have seen four spacecraft fly close to the planet and a fifth will soon become the first artificial Jovian satellite. Three of the same five spacecraft have travelled beyond Jupiter and visited the smaller, but otherwise similar, gaseous planet Saturn and one is continuing outward for a 1986 encounter with Uranus.

The atmosphere and space environment of Jupiter are so foreign, with experience of only the inner solar system to guide us, that the first missions to Jupiter were low-cost 'precursors', intended to explore the general environment – particularly with regard to the intense radiation belts surrounding Jupiter – before more exotic spacecraft were risked. These – Pioneers 10 and 11 – carried fairly simple visible, ultraviolet, and infrared sensors for preliminary remote sensing of the atmosphere. The ultraviolet instrument was just a two-channel photometer. The two channels are located at 1216 Å (the hydrogen Lyman-α line) and 584 Å, a resonant emission of helium. These are selected using filters. In the case of the helium channel, a thin film of aluminium acts as a filter which, when coupled with a lithium fluoride photocathode, defines a spectral pass band of approximately 200 to 800 Å. The hydrogen channel uses the front surface of the aluminium film as a photocathode obtaining a response out to 1400 Å. This detects both hydrogen and helium emissions, the former about ten times more intense. At Jupiter the instrument measured a hydrogen intensity of about 1000 rayleighs and a helium intensity in the range 10–20 rayleighs. This was the first detection of helium on Jupiter and permitted an estimate of the bulk composition of the planet: 85% hydrogen, and 15% helium. Within experimental error, this mixture corresponds to that which would be expected in a 'primordial' mixture, i.e. one made up of the elements in their cosmological proportions. Jupiter is large and cold enough to have retained its composition largely unmodified since the formation of the solar system. On the Earth, of course, most of the lighter atoms and molecules escaped long ago.

The Pioneer 10 infrared radiometer (Bender *et al.* 1976) was a two-channel device, using magnesium oxide as a filter for channel 1 ($14-25\,\mu m$) and barium fluoride for channel 2 ($30-56\,\mu m$). These channels were chosen to pro-

Fig. 16.1 Weighting functions for the Pioneer 10/11 Infrared Radiometer, as a functions of the zenith angle $\wp$ (angle between the direction of view and local vertical) (from Orton 1975).

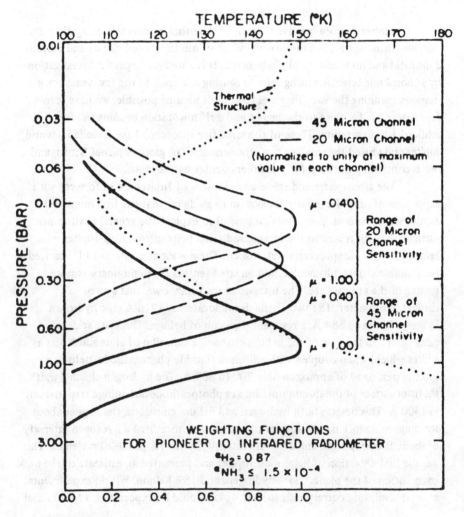

vide a good estimate of the total emitted flux from the planet, and also (from the ratio of the two emitted intensities) another estimate of the atmospheric hydrogen to helium ratio. A two-inch telescope with a 1° x 0.3° field of view and antimony-bismuth thermocouple detectors gave a 100:1 signal to noise ratio in both channels for a brightness temperature of 130 K. The instrument was used for temperature sounding by Orton *et al.* (1975) by using the angular

Fig. 16.2 Jovian temperature profile models retrieved from Pioneer 10 infrared radiometer data (from Orton 1975).

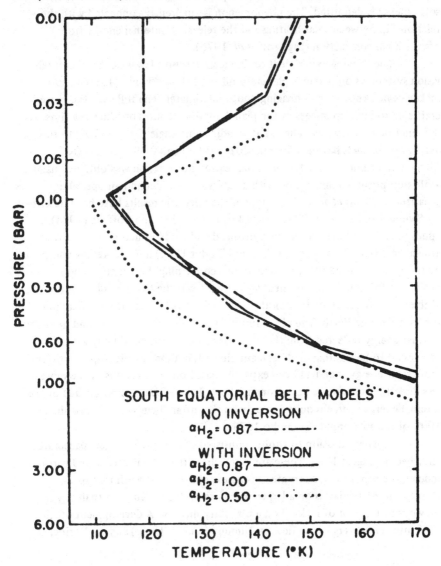

dependence of the weighting function to probe the atmosphere (Fig. 16.1). Some of the resulting temperature profiles are shown in Fig. 16.2. These are generally in accord with radiative equilibrium calculations, but show small but significant differences between belts (clear) and zones (cloudy). The measurements confirmed the Earth-based finding that Jupiter emits more energy than it receives: approximately twice as much, according to Pioneer 10/11 data (Ingersoll *et al.* 1976). Furthermore, and rather curiously, Jupiter radiates just as warmly at the poles as it does at the equatorial regions where most of the solar energy is deposited. The phenomenon is apparently connected with the relationship between solar heating and the release of internal energy from gravitational contraction (Ingersoll *et al.* 1976).

The imaging experiment on Pioneers 10 and 11 was not a vidicon television system, as on previous planetary missions, but a smaller, lighter photometric device known as an imaging photopolarimeter. The IPP used the spacecraft spin to built up images of the planet and its satellites one line at a time. A 0.5 mrad field of view was aligned at an adjustable angle (27° to 170°) to the spin axis. Two channels having $0.595-0.720\,\mu m$ (blue) and $0.39-0.50\,\mu m$ (red) filters used channeltron detectors. Polarization information was obtained using a Wollaston prism in conjunction with a half-wave retarder and an optical depolarizer (Pellicori *et al.* 1973). Some of the scientific results have been described by Swindell and Doose (1974); Tomasko *et al.* (1976), and Coffeen (1974). In addition to spectacular images of dynamical and cloud structure in the Jovian atmosphere (see, for example, Briggs and Taylor 1982) limb darkening and polarimetry have permitted tentative deductions concerning the structure and microphysics of the cloud. Also of great significance was the first measurement of the visible phase function of Jupiter at phase angles greater than the 11° maximum subtended at the Earth. The phase function describes the direction and intensity of solar energy reflected from the planet into space. The total integrated energy was found to be less than had been anticipated; in other words, Jupiter absorbs more solar energy than had been expected based on observations of the more-or-less directly backscattered component alone. This reduces the magnitude of the internal energy source required to explain the infrared emissions to less than 100% of the solar input (Tomasko 1976).

The next mission to Jupiter consisted of the two Voyager spacecraft launched in August 1977. These are much larger than the Pioneers, and carry much more sophisticated remote sensing experiments. The infrared spectrometer in particular, is on a larger and more sophisticated scale than any previous instrument of its kind on a planetary mission. A development of the IRIS interferometers which flew on Nimbus 3 and 4 (Ch. 6) and on Mariner 9

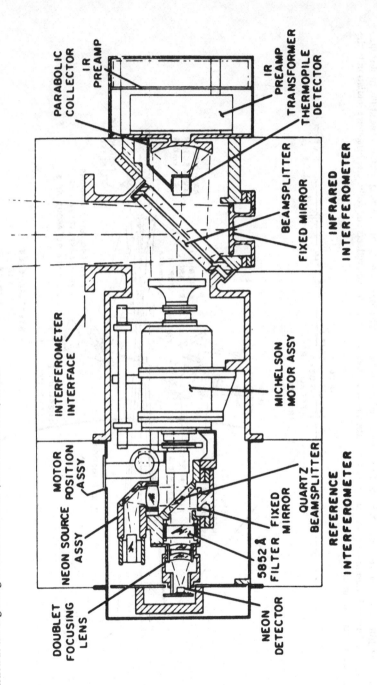

Fig. 16.3 The Voyager IRIS main IR and references interferometer configuration. Mirrors mounted on opposite ends of a common motor shaft couple the main and reference interferometers. The motor motion is phase locked to the spacecraft clock. The reference interferometer signal ensures data sampling at precise increments of mirror displacement. The motor scans in one direction and is returned to a predetermined start position at the beginning of each data frame (from Hanel et al., 1980).

292

Fig. 16.4 Optical layout of the Voyager IRIS. The primary and the secondary telescope mirrors form an image of the scene at the field stop. The spherical dichroic mirror transmits short-wave radiation to the radiometer detector and reflects and collimates long-wave radiation to the IR interferometer (from Hanel et al., 1980).

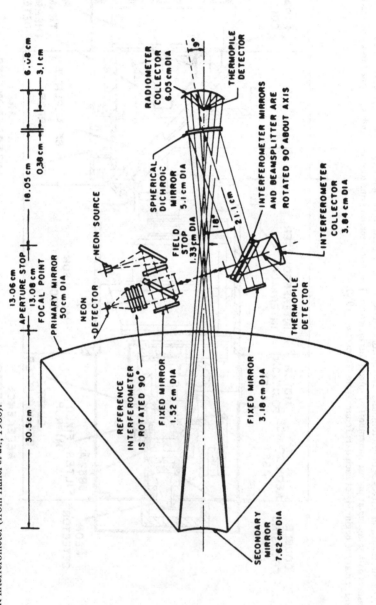

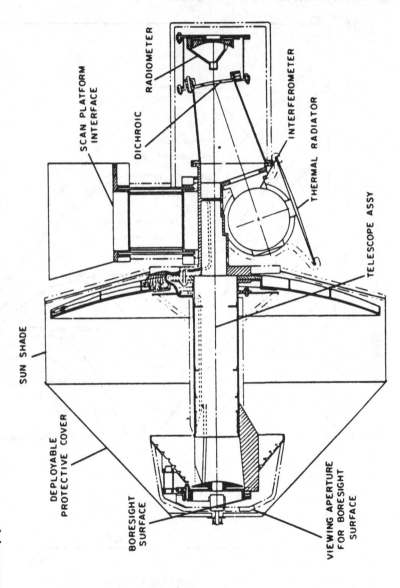

Fig. 16.5 Configuration of the Voyager IRIS instrument. The instrument is mounted on the spacecraft scan platform as shown. The protective cover was jettisoned several days after launch. In this view, the IR interferometer is seen end-on; i.e., the motion of the Michelson mirror is perpendicular to the paper. The reference interferometer is directly behind the IR interferometer (from Hanel *et al.*, 1980).

Fig. 16.6 The intensity of emitted and reflected energy as a function of wavelength for simple models of the planets. The real spectra are modified by atmospheric absorption and emission features but have this general form.

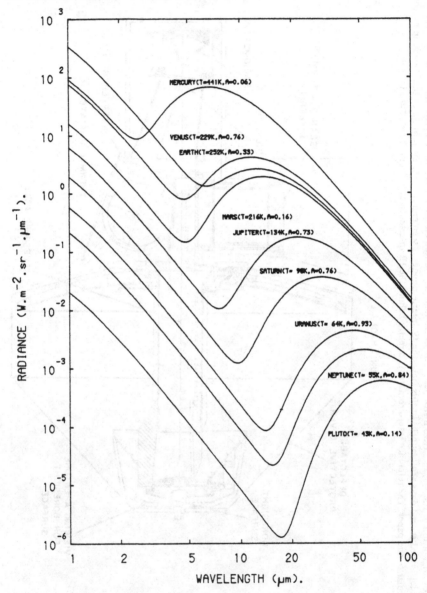

to Mars (Ch. 14), its physical appearance is dominated by a large telescope mirror (Figs. 16.3, 4, 5). Actually 50 cm in diameter, this is required to obtain adequate signal to noise ratios for the low radiant energy levels at Jupiter and Saturn. The outer planets are so far from the sun (778 x 10^6 km for Jupiter and 1426 x 10^6 km for Saturn) that both the reflected solar intensity in the visible, ultraviolet and near-infrared and the thermally emitted intensity are more than an order of magnitude lower than for the Earth. This is dramatically illustrated in Fig. 16.6, where we show the reflected and emitted energy as a function of wavelength for five of the planets. These are simplified spectra corresponding to grey bodies at the mean temperatures and albedos of the planets. The detailed structure due to molecular vibration rotation bands and wavelength-dependent cloud and surface properties are small on this scale, in spite of the fact that they contain most of the information on which the remote sensing experiment are based.

The atmosphere of Jupiter contains no carbon dioxide, and so vertical temperature sounding calls for a different approach to that which works on Venus, Earth and Mars. The hydrogen and helium molecules, which make up over 99% of the Jovian atmosphere, have no permanent dipole moments and hence are highly transparent to thermal infrared radiation at low pressures. At higher pressures, hydrogen develops a strong 'continuum' (i.e. slowly varying with frequency) type absorption spectrum, and it is this, together with the clouds, which provides the major source of opacity from about 15 to about 50 μm wavelength. This 'pressure induced' hydrogen spectrum is caused by collisional distortion of the molecules inducing a temporary dipole moment, and the intensity of the absorption and its shape is modified by collisions with helium atoms which themselves neither absorb nor emit in the infrared. In addition to hydrogen and helium, the Jovian atmosphere contains a great deal of methane (about 0.4% by volume) and ammonia (about 0.1%), each of which has a rich vibration–rotation spectrum.

The pure rotational band of ammonia is the dominant source of opacity on Jupiter from about 50 μm out to radio wavelengths. At the shorter wavelength end of the thermal spectrum ammonia again dominates from about 9–15 μm by virtue of its ν_2 band, while the ν_4 band of methane fills the 7–9 μm region. Below 7 μm the Jovian spectrum is no longer dominated by thermal radiation (Fig. 16.6), except for a small transparent 'window' region near 5 μm. The question of which part of the spectrum from 7–1000 μm to employ as the Jovian analogue of the 15 μm CO_2 band, for temperature sounding purposes, is addressed in Fig. 16.7. There it is seen that, looking in any part of the spectrum dominated by hydrogen or ammonia, one 'sees' to almost the same level. The reason is easy to find – pressure-induced absorption depends on the square of

296

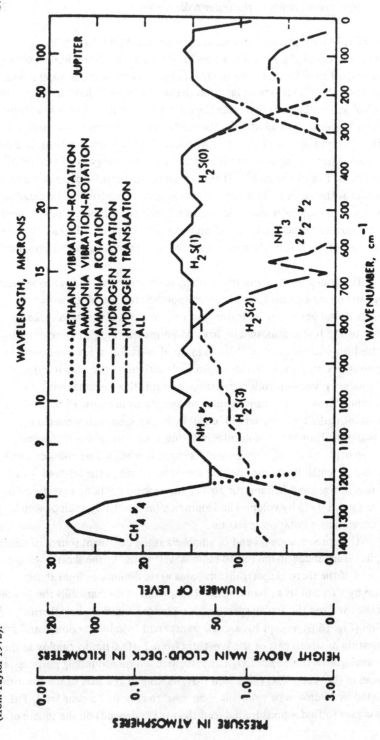

Fig. 16.7 The level at which optical depth is unity, calculated as a function of wavelength for a model Jovian atmosphere without cloud (from Taylor 1972).

the pressure and hence the atmosphere, once it starts to absorb, increases its opacity to totally absorbing over a relatively small height range, almost independent of the absorption coefficient itself.

Somewhat by coincidence, ammonia saturates on Jupiter at about the same level where hydrogen becomes opaque, and so is greatly depleted at higher levels. (In this respect, ammonia on Jupiter corresponds to water on Earth and behaves in an analogous manner with regard to cloud formation, latent heat transportation and so on.) This leaves methane as the only species present in large quantities which is uniformly mixed and which offers a large range of opacities as needed for temperature sounding. Fig. 16.8 shows a set of weighting functions calculated for the ν_4 band. As a corollary to these, it is seen from Fig. 16.7

Fig. 16.8 Weighting functions for six spectral bands, centred on the wavenumbers shown, in the ν_4 band of methane at 7.7 μm. A model Saturnian temperature-pressure profile is also shown (from Taylor 1972).

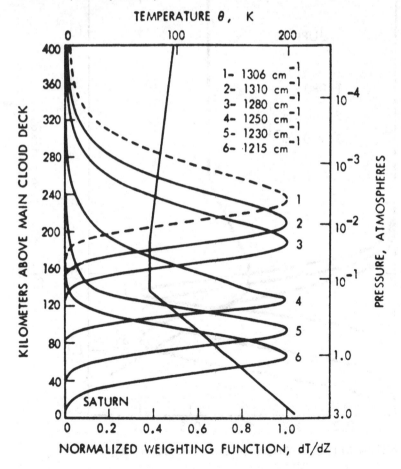

that a limited vertical coverage is possible by observing in the far infrared near
40 μm, in the region between the edges of the hydrogen and ammonia bands.
The weighting functions corresponding to this are shown in Fig. 16.9. The wavelength coverage of the Voyager IRIS is such (2.5–50 μm) that temperature
sounding in both of these regions is possible. Simultaneous sounding in two
widely different wavelength bands is a powerful technique for avoiding mis-
leading conclusions due to the unknown opacity effects of thin clouds or aerosols,
since the latter will normally be a strong function of wavelength. Fig. 16.10
shows some examples of retrieved temperature profiles using this technique and

Fig. 16.9 As Fig. 16.8, but for three channels in the far infrared.

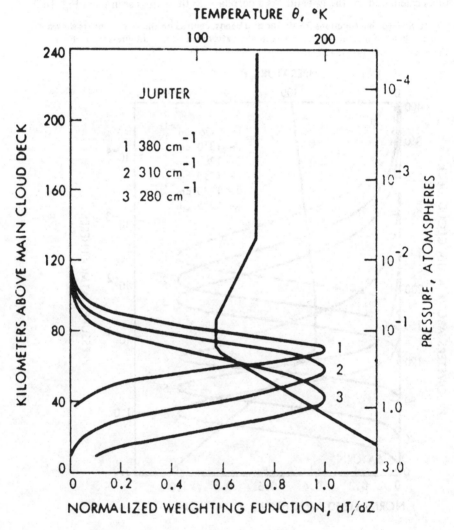

Fig. 16.11 the single profile obtained by the radio occultation method during the passage of Voyager 1 behind Jupiter. The two very different techniques produce results which are in very good general agreement even when applied to the alien atmospheric conditions found on Jupiter.

The great advantage of infrared remote sensing is its coverage; as an example of this, Fig. 16.12 shows details of the temperature structure in the Great Red Spot as determined from IRIS soundings. A further example shows how global coverage may be applied to understanding the dynamics of the banded cloud structure. Fig. 16.13 compares the velocity at a particular pressure level (152 mbar) as determined by the IRIS and by the Voyager imaging system. The former obtains winds by inserting retrieved temperatures into the thermal wind equation (Pirraglia *et al.* 1981) while the latter uses cloud-feature tracking.

Fig. 16.10 Examples of Jovian temperature profiles obtained by inverting radiances at the frequencies of Fig. 16.8 and 16.9, measured by the Voyager IRIS instruments (from Hanel *et al.* 1979).

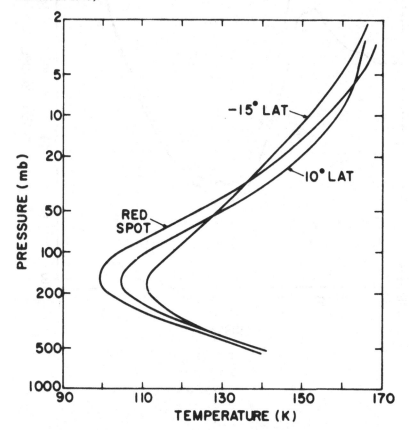

300

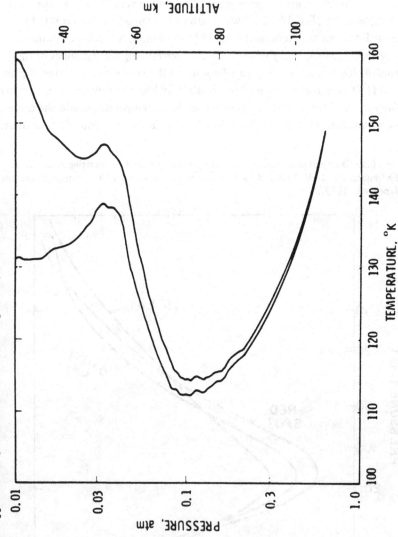

Fig. 16.11 Temperature profile at 12°S latitude measured by the Voyager 1 radio occulation experiment. The two curves show the effect of assuming different upper boundary conditions in the inversion process (from Eshleman *et al.* 1979).

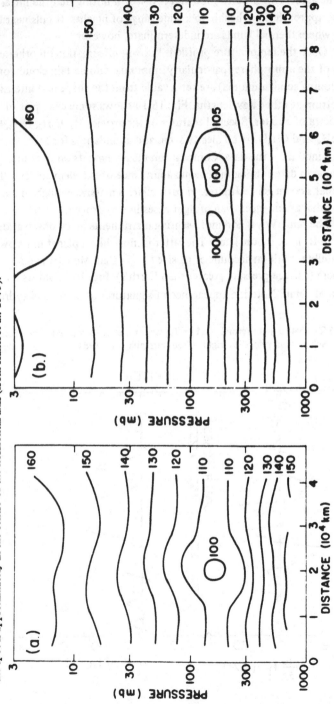

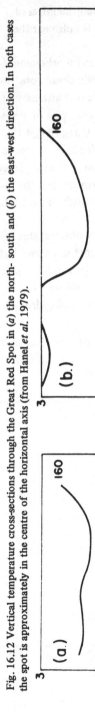

Fig. 16.12 Vertical temperature cross-sections through the Great Red Spot in (*a*) the north- south and (*b*) the east-west direction. In both cases the spot is approximately in the centre of the horizontal axis (from Hanel *et al.* 1979).

Again, the agreement is quite good and it is worth noting that the infrared sounding approach is applicable to a wide range of heights. It fails near the equator where thermal winds are indeterminate, however.

Once the temperature profile is known, information on other components of the atmosphere (particularly ammonia, and certain cloud constituents including solid ammonia) are retrievable from the shape and intensity of the spectrum at other wavelengths. Fig. 16.14 shows some examples of Voyager IRIS spectra of Jupiter. Spectral features of the species H_2, C_2H_2, NH_3, CH_4, H_2O, GeH_4 and CH_3D are all clearly seen, and abundances for these species can be determined from the data. Composition measurements on the outer planets are of value in determining the cosmic abundance of the elements and the history of the solar system, since Jupiter, in particular, is massive enough to have retained its primordial abundances even of light gases like hydrogen.

The same Voyager remote-sensing instruments made observations of Saturn and its large moon Titan. The latter is more like a planet in its own right than a satellite, both on account of its size (larger than Mercury) and its thick atmosphere (surface pressure greater than Earth's). Fig. 16.15 shows an IRIS spectrum of Titan. Note the appearance of signatures of numerous hydrocarbon

Fig. 16.13 The calculated thermal wind on Jupiter at the cloud top level (left) compared to cloud-tracked wind velocities (right) (from Pirraglia *et al.* 1981).

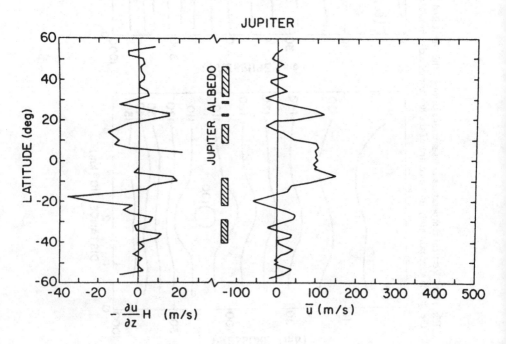

Fig. 16.14 Thermal emission spectra for Voyager 1 and 2. Strong spectral features of the gases H_2, C_2H_2, NH_3, CH_4 appear. Significant PH_3 absorption occurs in the region between 1100 and 1200 cm^{-1}, the strongest feature being the Q branch at 1122cm^{-1} (right). Thermal emission spectra in the 2000 cm^{-1} Jovian window for Voyager 2. Strong absorption features of NH_3, H_2O, GeH_4 and CH_2D appear. (from Hanel *et al.* 1979).

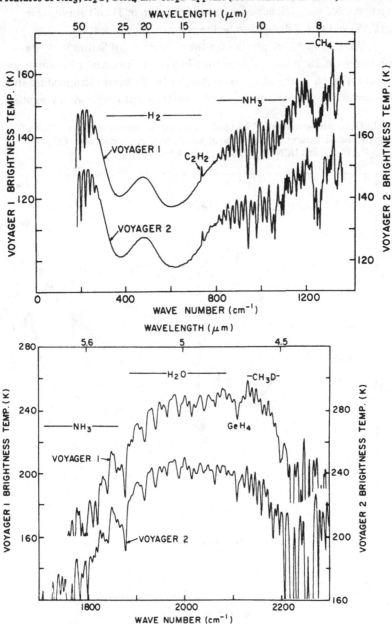

species and also hydrogen cyanide, the latter in amounts of less than one part per billion. Fig. 16.16 shows the Titanian vertical temperature profile obtained from a combination of IRIS and radio occultation results, with the probable locations of cloud and haze layers indicated. The ultraviolet spectrometer on Voyager revealed the presence of molecular nitrogen in Titan's atmosphere (Fig. 16.17) and this is thought to be the principal constituent.

Some temperature profiles for the atmosphere of Saturn itself are shown in Fig. 16.18. Again, infrared sounding and radio occultation are in good agreement. Fig. 16.19 shows a cross-section of the Saturnian temperature structure as measured by IRIS; conversion of these temperatures to wind velocities

Fig. 16.15 Equatorial spectrum of Titan and calculated comparison spectrum. Most of the strong emission features are due to the minor stratospheric constituents CH_4, C_2H_2, C_2H_4, C_2H_6, and HCN (from Hanel *et al.* 1979).

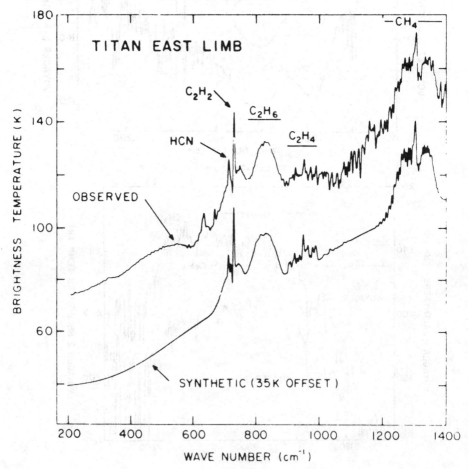

Fig. 16.16 Temperature profile of Titan. The solid curves are from IRIS data and the dashed parts are from the profile of the temperature-molecular weight ratio obtained by the Voyager radio science investigation. Haze and condensation zones are indicated schematically (from Hanel *et al.* 1979).

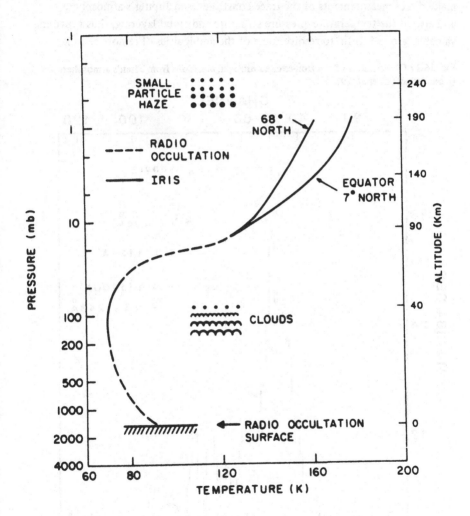

results in much poorer agreement with cloud-tracked winds than for Jupiter, however, particularly at low latitudes (Fig. 16.20).

After Voyager, the next mission to the outer planets will be a Jupiter orbiter-entry probe combination known as Galileo. The orbiter will occupy a highly eccentric, distant orbit to avoid the Jovian radiation belts and also to permit several close encounters of the Galilean satellites. The entry probe will make direct measurements of the trace constituents in Jupiter's atmosphere, and also of the temperature-pressure structure and cloud layering, thus providing valuable 'ground truth' to resolve some of the ambiguities of remote sensing.

Fig. 16.17 Observations of electron-excited nitrogen emissions from Titan's atmosphere (from Broadfoot *et al.* 1981).

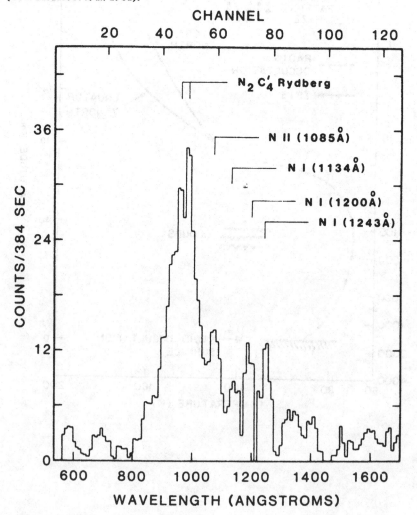

The Galileo orbiter payload includes a near-infrared spectrometer, a combined visible and near-visible photopolarimeter and far-infrared radiometer, an ultraviolet spectrometer and a television experiment. The latter uses a solid-state camera, rather than the more usual videocon, for the first time. A particular advantage of this approach is a spectral response which extends out to 1.1 μm in the infrared, which gives better discrimination of different cloud types and heights. The ultraviolet spectrometer is of the Ebert–Fastie type, covering the range from 0.115 to 0.430 μm with 0.001 μm of resolution. Its objectives are the stratospheric abundances of ammonia and hydrocarbon molecules, the exospheric abundance and distribution of atomic and molecular hydrogen, and the hydrogen to helium abundance ratio.

Fig. 16.18 Vertical temperature profiles for Saturn and Jupiter, as retrieved by inversion of Voyager IRIS spectra. The dashed portions of the curves represent extrapolations along adiabats (from Hanel *et al.* 1981).

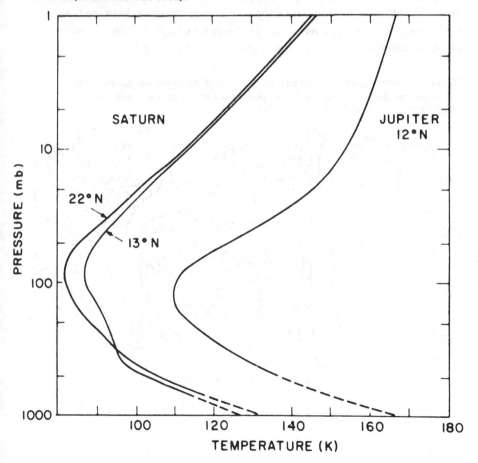

Fig. 16.21 shows the photopolarimeter-radiometer (PPR) and Table
16.1 its spectral channels. With these it will investigate the cloud microstructure
by measuring the solar phase angle dependence of the polarization of reflected
light, and the temperature structure in a limited vertical range above the clouds
by far-infrared emission measurements. A particularly innovative instrument is
the Near-Infrared Mapping Spectrometer (NIMS); this uses a rocking grating and
a separately driven rocking mirror to scan the target in the spectral and spatial
dimensions simultaneously (Fig. 16.22). With sensitive detectors cooled to the
temperature of liquid nitrogen by a cooler which radiates heat away to space,
the NIMS rapidly builds up an image of the area in view in which each picture
element contains a spectrum from $0.7-5.2\,\mu$m. These 'three-dimensional' images
will be used to study the complex layering and scattering properties of the
Jovian clouds, the spatial variability of condensible minor constituents, like
ammonia and phosphine, the run of temperature in the 1–5 atm pressure range
(where not obscured by clouds) and the atmospheric dynamics. Between them,
NIMS and PPR will cover a vertical range of seventy or eighty kilometres within
which the temperature varies from just above 100 K to nearly 300 K.

Fig. 16.19 Thermal structure of the atmosphere in the N. hemisphere of Saturn, with
the average profile subtracted, and the positions of the main belts and zones indicated
along the top edge (from Pirreglia *et al.* 1981).

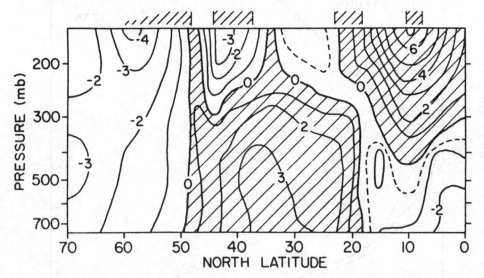

Table 16.1 *PPR spectral bands*

Function	Centre wavelength, λ_c	FWHM,* $\Delta\lambda$	F/R position(s)
Photopolarimetry	410.0 ± 4 nm	40 ± 6 nm	13, 15, 17
	677.7 ± 2	9 ± 2	7, 9, 11,
	945.0 ± 3	10 ± 2	1, 3, 5,
Photometry	618.0 ± 2 nm	9 ± 2 nm	25
	633.0 ± 2	10 ± 2	26
	646.0 ± 2	8 ± 2	27
	789.0 ± 3	12 ± 3	28
	829.7 ± 2	10 ± 2	29
	841.0 ± 2	7 ± 2	30
	891.7 ± 2	16 ± 3	31
Radiometry	Solar plus thermal	Open	24
	Solar	$0.3\,\mu m \lesssim \lambda \lesssim 4\,\mu m$	23
	17 μm	4 ± 1 μm	18
	21 μm	4 ± 1	19
	27.5 μm	9 ± 2	21
	37 μm	10 ± 3	20
	$\lambda \gtrsim 42 \pm 3\,\mu m$	NA	22

*Full width at half maximum

Fig. 16.20 A comparison between thermal winds, calculated from IRIS temperature soundings, and cloud-tracked winds on Saturn (from Pirraglia *et al.* 1981).

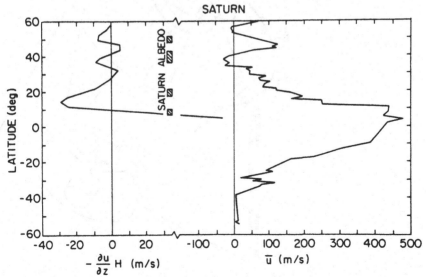

310

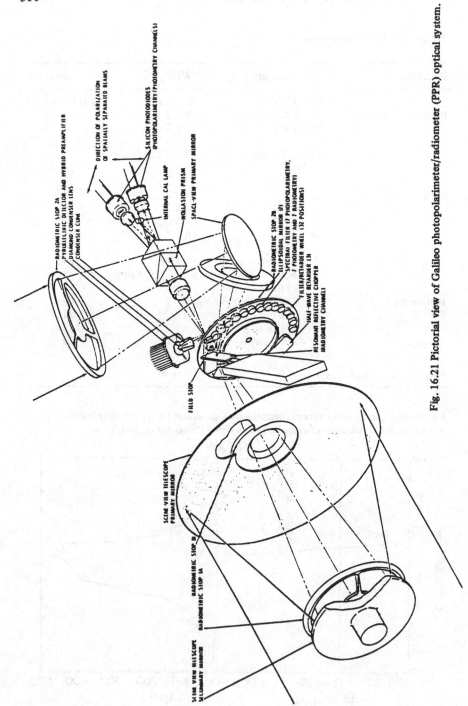

Fig. 16.21 Pictorial view of Galileo photopolarimeter/radiometer (PPR) optical system.

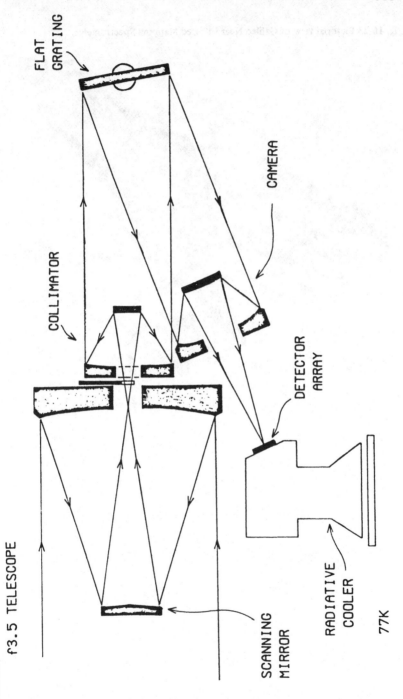

Fig. 16.22 Layout of the basic components of the Galileo Near Infrared Mapping Spectrometer.

Fig. 16.23 Pictorial view of Galileo Near Infrared Mapping Spectrometer.

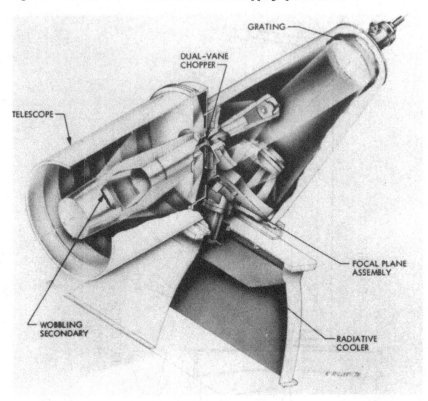

17

FURTHER DEVELOPMENTS

Remote sensing of atmospheric parameters, as a tool for research and exploration, has enjoyed a period of rapid growth which has lasted for nearly twenty years. At least as far as the technology is concerned, there are no signs of an end to this advancement on all fronts. In particular, new vehicles for launching larger, more numerous, and more sophisticated instruments, new technology and new techniques within the instruments themselves, radically improved ground-data processing hardware and missions to the most distant, unexplored planets are all visible on the horizon. Let us speculate briefly on what lies in store in each area.

The space vehicle, in particular the Earth satellite, has been revolution-ized by the advent of the space shuttle. With a payload of 32 000 lb, compared to the 1000 lb mass of a Nimbus satellite for example, and frequent launches, the size and number of remote-sensing experiments in orbit can increase dramatically within existing national budgets. It then becomes feasible, for example, to orbit instruments with very large optics, capable of much higher spatial resolution than before. Signal-to-noise ratios can be enhanced by virtue of the relative ease with which cryogenic devices for cooled detectors can be implemented. The use of large detector arrays, consisting of say 1000 elements or more, is now feasible thanks largely to the impetus supplied by the military of several nations. These allow good coverage to be enjoyed at the same time as the other benefits cited above, particularly since many identical instruments can be in orbit simul-taneously.

Among the many new or imminent advances in instrumental techniques for remote sensing, probably the capability for imaging at wavelengths where previously only single detector measurements were possible, is outstanding. Other examples are three-dimensional imaging in the visible and infrared, to obtain cloud heights, and multispectral imaging to obtain temperature and compositional structure from complete weather systems. A completely new infrared technique, currently in its infancy but exciting in its potential, is the measurement of the atmospheric wind field by measuring Doppler shifts in the emission or absorption features of the atmosphere at the limb. This seems feasible by either very high spectral resolution observations of a single line (using Fabry–Perôt etalons or laser heterodyne techniques, for example) or by a correlation method such as pressure modulator radiometry. The time cannot be too far away when the entire globe is surveyed continuously, at sufficiently high resolution both vertically and horizontally, with regard to all of the important meteorological parameters. Contemplating such a situation brings us face to face with the entire question of predictability – how many measurements is enough or, conversely, how much can be done with an essentially infinite amount of data? For the problem which is perhaps the most basic – weather forecasting – it seems likely that predictions can be made not more than ten to twenty days ahead with useful accuracy. Clearly this is much better than is achieved at the present, so there do not seem to be any fundamental barriers to progress for the next ten years at least. Less ambitious atmospheric problems, such as early storm detection and basic climatology, certainly will be solved by more modest advances in the type and quantity of observations from space.

To implement and take advantage of such advances, a parallel improvement in ground data handling capabilities is implied. Even at present, manipulating the enormous arrays of numbers from a single flight experiment in order to extract their information content is a daunting task. Fortunately, microelectronics capable of enormously large numbers of fast digital operations are equal to the task, and are progressing as fast as other parts of the field. Modestly-sized 'mini-computers' are capable of data-handling operations which would have required a large room full of much more expensive equipment only ten years ago. It seems, then, that there are no obvious impediments here, either, to a remote sensing capability which can continue to expand and tackle terrestrial problems, the surfaces of which we have previously only scratched.

We have seen in Chapter 8 that improving the coverage and quality of the basic data and the means of its assimilation has led to a significantly improved ability to forecast global weather patterns over periods of 5 to 10 days ahead. It is still a matter of debate how far further improvements may take us in the ability to forecast detailed weather patterns. Further work with the data set from the First GARP Global Experiment in 1979 will assist in assessing the extent to which detailed forecasting can be extended; the current wisdom is that two to three weeks will be the limit.

However, even if the details of weather systems are not predictable for more than weeks ahead, the question may be asked as to whether the average weather (often called the climate) can be predicted over larger periods of months, years or decades. To study the climate question the World Climate Research Programme (WCRP) has been set up as a joint enterprise between the International

Council of Scientific Unions (ICSU) and the World Meteorological Organisation (WMO); the programme is organised in a similar way to its predecessor, the Global Atmospheric Research Programme (GARP). The dual purpose of the WCRP is to determine (i) to what extent climate can be predicted and (ii) the extent of man's influence on climate. An important example of (ii) is the change in climate, both global and regional which could occur as a result of the increase of carbon dioxide due to man's burning of fossil fuels.

Consideration of the climate problem requires observation and understanding of all components of the climate system, namely the ocean (including sea ice), cryosphere, and land surface in addition to the atmosphere, together with the interactions between the various components (Houghton 1981 and Houghton (ed.) 1983). Remote sounding measurements will play a large part in getting the information required not only for the atmosphere but also for the other components. Regarding the oceans, we have already referred to the remote sounding of sea surface temperature (para. 3.2) and to surface stress (para. 12.2). Another important measurement is that of surface topography by radar altimetry from which the motion of the ocean in its surface layers can be determined. Very high accuracy of the order of a few cm in surface elevation is required, but measurements from Seasat have shown that such an accuracy is possible. The same technique can be used to give information about the changes which may be occurring in the major ice sheets. Regarding the land surface, and important properties such as soil moisture, albedo and vegetation cover, observations (active and passive) from a wide range of wavelengths from the visible through the infrared to the microwave can in principle provide most of the information required although a great deal of progress needs to be made in methods of data interpretation if the needs of climate research are going to be met.

We have seen in earlier chapters how the principles of remote sensing have been applied to planetary atmospheres other than our own. Most often, successful Earth satellite techniques have been modified to make them adapt to the atmospheric conditions elsewhere. Occasionally, totally new techniques have been devised for problems or conditions which have no terrestrial parallel. We can expect this process to go on; high-energy upper stages for the space shuttle offer the promise of larger and more numerous spacecraft for planetary missions than before. Thus we can expect to have missions to the most distant planets as soon as political expediency sees fit. Saturn and Uranus orbiters, and Neptune fly-bys, have already been the subjects of detailed studies. More exciting still is the prospect that the nearer planets, especially Mars and Venus of the 'terrestrial' group, will be the subject of detailed meteorological coverage by networks of satellites and sondes similar to those covering the Earth at present. Such a situation will mark the beginning of large-scale comparative planetary meteorology.

REFERENCES

Aanensen, C. J. M. 1973 The use of Nimbus 4 radiance and radiosonde data in the construction of stratospheric contour charts. *Quart. J. R. Met. Soc.* 99, 657-668

Abel, P. G. *et al.* 1970 Remote sounding of atmospheric temperature from satellites. II. The Selective Chopper Radiometer for Nimbus D. *Proc. R. Soc. Lond.* A320, 35-55

Adler, R. F. 1975 Mean meridional circulation in the southern hemisphere stratosphere based on satellite information. *J. Atmos. Sci.* 32, 893-898

Akosofu, S. I. 1976. Recent progress in studies of DMSP Auroral photographs. *Space Science Rev.* 19, 169-215

Akvilonova, A. B., Kutulza, A. G. and Mitnik, L. M. 1971 *Isv. Atmos. Ocean. Phys.* 7, 133-138.

Allison, L. J. Schumugge, T. J. & Byrne, G. 1979 A hydrological analysis of East Australia floods using Nimbus 5 ESMR data. *Bull. Amer. Met. Soc.* 60, 1414-1427

Anderson, G. P., C. A. Barth, F. Cayla, and J. London, 1969 Satellite observations of the vertical ozone distribution in the upper stratosphere. *Ann. Geophys.* 25, 239- 243

Andrews, D. G. and McIntyre, M. E. 1976 Planetary waves in horizontal and vertical shear: the generalized Eliassen-Palm relation and the mean zonal acceleration. *J. Atmos. Sci,* 33, 2031-2048

Asgeirsson, V. and Stanford, J. L. 1977 Systematic deviations of Nimbus 5 atmospheric temperature fields from radiosonde data over the winter antarctic. *Geophys. Res. Lett.* 4, 445-447

Astheimer, R. W., De Waard, R. and Jackson, E. A. 1961 *J. Opt. Soc. Amer.* 51, 1386-1393.

Austen, M. D. *et al.* 1977 Satellite temperature measurements in the 40-90 km region by the Pressure Modulator Radiometer. *COSPAR Space Research* XVII, 111-115

Austen, M. D. *et al.* 1976 Satellite observations of planetary waves in the mesosphere. *Nature,* 260, 594-596

Backus, G. E. and Gilbert, J. F. 1970 Uniqueness in the inversion of inaccurate gross earth data. *Phil. Trans. Roy. Soc. Lond.* A266, 123-192

Baer, F. 1977 Adjustment of initial conditions required to suppress gravity oscillations in non-linear flows. *Contrib. Atmos. Phys.* 50, 350-366.

Bailey, P. L. and Gille J. C. 1978 in *Remote sensing of the atmosphere: inversion methods and applications.* (Eds A. L. Fymat and V. E. Zuev) 115-122. Amsterdam: Elsevier

Barath. F. *et al.* 1964 Mariner 2 microwave experiment and results, *Ap. J,* 69, 49-58

Barnett, T. L., 1969 Application of a nonlinear least squares method to atmospheric temperature sounding. *J. Atmos. Sci.* 26, 457

Barnett, J. J. *et al.* 1972 The first year of the selective chopper radiometer on Nimbus 4. *Quart. J. R. Met. Soc.* 98, 17-37

Barnett, J. J. and Walshaw, C. D. 1972 Temperature measurements from a satellite: applications and achievements. *Environmental remote sensing,* 187-213. (Eds Barrett and Curtis). Arnold.

Barnett, J. J. 1973 Analysis of Stratospheric measurements by the Nimbus IV and V selective chopper radiometer. *Quart. J. R. Met. Soc.* 99, 173-188

Barnett, J. J. *et al.* 1973 Stratospheric observations from Nimbus 5. *Nature* 245, 141-143

Barnett, J. J. 1974 The mean meridional temperature behaviour of the stratosphere from November 1970 to November 1971 derived from measurements by the selective chopper radiometer on Nimbus IV. *Quart. J. R. Met. Soc.* 100, 505-530

Barnett, J. J. 1975 Hemispheric coupling - evidence of a cross-equatorial planetary waveguide in the stratosphere. *Quart. J. R. Met. Soc.* 101, 835-845

Barnett, J. J. 1975 Large sudden warming in the southern hemisphere. *Nature,* 255, 387-389

Barnett, J. J. *et al.* 1975 Comparison between radiosonde, rocketsonde and satellite observations of atmospheric temperatures. *Quart. J. R. Met. Soc.* 101, 423-436

Barnett, J. J., Houghton T. and Pyle J. A. 1975 The temperature dependence of the ozone concentration near the stratospause, *Quart. J. R. Met. Soc.* 101, 245-257

Barnett, J. J. 1977 The antarctic atmosphere as seen by satellites. *Phil. Trans. R. Soc. Lond.* 279, 247-259

Barnett, J. J., Alyea, F. N. and Cunnold, D. M. 1977 Comparison between satellite radiance observations and those derived from a stratospheric numerical model. *Proceedings of IAGA/IAMAP Joint Assembly,* Seattle.

Barnett, J. J. and Crane, A. J. 1977 Energetics of the upper stratosphere during a sudden warming. *Proceedings of IAGA/IAMAP Joint Assembly,* Seattle

Barnett, J. J., Houghton, J. T. and Peskett, G. D. 1978 *Proceedings of COSPAR symposium,* Innsbruck.

Barth, C. A. 1981 Solar Mesosphere Explorer to study ozone. *Nature* 293, 259-260

Barth, C. A. *et al.* 1969 Mariner 6: The ultraviolet spectrum of Mars upper atmosphere. *Science* 165, 1004-1005

Barth, C. A. *et al.* 1971 Mariner 6 and 7 ultraviolet spectrometer experiment: upper atmosphere data, *J. Geophys. Res.* 76, 2213

Barth, C. A. *et al.* 1972 Mariner 9 UV spectrometer experiment: Mars airglow spectroscopy and variations in Lyman Alpha. *Icarus* 17, 457

Basharinov, A. E., Gorelik, A. G., Kalashnikov, V. V. and Kutuza, B. G. 1979 Simultaneous radiometric and radar measurements of the meteorological parameters of clouds and rain. *Izvestiya, Atmospheric and Oceanic Physics* 6, 301-304

Basharinov, A. Y., Gurvich, A. S., and Yegorov S. T. 1969 *Dokl. Akad. Nauk. USSR* 188, 1273-6 Sov. Phys. Dokl, English translation

Belton, M. J. S. *et al.* 1976 Cloud patterns, waves and convection in the Venus atmosphere. *J. Atmos. Sci.* 33, 1394-1417

Bender, M. L. *et al.* 1976 Infrared radiometer for the Pioneer 10 and 11 missions to Jupiter. *Applied Optics,* 13, 2623-2628

Bengtsson, L. and Kallberg, P. 1981 Numerical simulation - assessment of FGGE data with regard to their assimilation in a global data set. *I.C.S.U. Commission for Space Res.,* 1, No. 4, Oxford, 167-187

Bengtsson, L. 1979 Problems of using satellite information in numerical weather prediction. Proc. Tech. Conference on *'Use of data from meteorological satellites'* Lannion, France, 17-21 Sept. 1979 (ESA SP-143, October 1979).

Bertaux, J. L., Blamont, J. E., Lepine, V. M., Kurt, V. G., Romanova, N. N. and Smirnov, A. S. 1981 Venera 11 and 12 observations of EUV emissions from the upper atmosphere of Venus. *Planet. and Sp. Sci.,* **29,** 149-166

Berthorsson, P. and Döös, B. R. 1955 Numerical weather map analysis. *Tellus,* **7,** 329-340

Bishopf, W. and Bolin, B. 1966 *Tellus* **18,** 155-9

Blamont, J. E. and Luton, J. M. 1972 *Geophys. Res.* **77,** 3534-56

Bonner, W. P., Lemar, R., Van Haaren, A., Desmarais and O'Neil, H. 1976 A test of the impact of NOAA-2 VTPR soundings on operational analysis and forecasts, *NOAA Tech. Memo. NWS-57,* National Weather Service Spring, Md.

Bowen, R. A., Fusco, L., Morgan, J. and Roska, M. O. 1979 Operational production of cloud motion vectors (satellite winds) from Meteosat image data. In *'Use of data from meteorological satellites'*. Publ. SP-143 European Space Agency, Paris 65-75

Boyd, J. P. 1976 The non-interaction of waves with the zonally averaged flow on a spherical earth and the inter-relationships of eddy fluxes of energy, heat and momentum. *J. Atmos. Sci.* **33,** 2285-2291

Briggs, G. A. *et al.* 1977 Martian dynamical phenomena during June-Nov. 1976: Viking Orbiter imaging results. *J. Geophys, Res.* **82**

Briggs, G. A. and Taylor, F. W. 1982 *Cambridge Photographic Atlas of the Planets.* Cambridge University Press.

Broadfoot, L. *et al.* 1974 Mercury's atmosphere from Mariner 10: Preliminary results. *Science* **185,** 166-169

Broadfoot, L. and Shemansky, 1975 *The helium atmosphere of Mercury.* Paper presented at Fall meeting of the American Geophysical Union

Broadfoot, L. *et al.* 1977 Mariner 10 ultraviolet spectrometer: *Sp. Sci. Inst.* **3,** 199

Budden, K. G. 1961 *Radio Waves in the ionosphere.* London: Cambridge University Press

Burch, D. E. 1970 Investigation of the absorption of infra-red radiation by atmospheric gases. *Publ. 4-4784 Philco-Ford Corp.* Aeronautronic Div. Newport Beach, California

Burke, W. J. Schmugge, T., Paris, J. F. 1979 Comparison of 2.8 and 21 cm microwave radiometer observations over soils with emission model calculations. *J. Geophys. Res.* **84,** 287-294

Callis, L. B. and Nealy, J. E. 1979 Solar u.v. variability and its effect on stratospheric thermal structure and trace constituents, *Geophys, Res. Letters.* **5,** 249-252

Carr, M. *et al.* 1972 Imaging experiment; The Viking Mars Orbiter. *Icarus* **16,** 17-33

Chahine, M. T. 1968 Determination of the temperature profile in an atmosphere from its outgoing radiance. *J. Opt. Soc. Am.* **58,** 1634

Chahine, M. T. 1970 Inverse problems in radiative transfer: a determination of atmospheric parameters. *J. Atmos, Sci* **27,** 960

Chahine, M. T. 1972 A general relaxation method for inverse solution of the full radiative transfer equation. *J. Atmos Sci* **29,** 741

Chahine, M. T. 1976 Remote soundings of cloudy atmosphers, 1, The single cloud layer. *J. Atmos Sci.* **31,** 233-243

Chapman, W. *et al.* 1974 A spectral analysis of global atmospheric temperature fields observed by selective chopper radiometer on the Nimbus 4 satellite during the year 1970-1. *Proc. R. Soc. Lond.* A338, 57-76

Chapman, W. A. and McGregor, J. 1978 The application of complex demodulation to meteorological satellite data. *Quart. J. R. Met. Soc.* **104,** 213-223

Charney, J. G. and Drazin P. G. 1961 Propagation of planetary-scale disturbances from the lower to the upper atmosphere. *J. Geophys. Res.* **66,** 83-109

Chase, S. T. and Neugebauer, G. 1963 The Mariner 2 infrared radiometer experiment. *J. Geophys. Res.* **68,** 6157

Chase, S. T. *et al*. 1974 Preliminary infrared radiometry of Venus from Mariner 10. *Science* 183, 1291-1292

Chen, E. *et al*. 1979 Satellite-sensed winter nocturnal temperature patterns of the everglades agricultural area. *J. Appl. Met*. 18, 992-1002

Chow, M. D. 1974 An iterative scheme for determining sea surface temperature, temperature profiles, and humidity profiles from satellite-measured infrared data. *J. Geophys. Res* 79, 430-434

Chu, W. P. and McCormick, M. P. 1979 Inversion of stratospheric aerosol and gaseous constituents from solar extinction data. *Appl. Opt*. 18, 1404-1413

CIRA 1972 *COSPAR International reference atmosphere 1972*, Akademie-Verlag, Berlin

Coffeen, D. L. 1974 Optical polarization measurements of Jupiter at 1030 phase angle. *J. Geophys. Res*. 79, 3645-3652

Coffey, M. T. 1977 Water vapour absorption in the 10-12 μm atmospheric window. *Quart. J. R. Met. Soc*. 103, 685-692

Colin, L. M. and Hunten, D. M. 1977 Pioneer Venus Experiment descriptions. *Sp. Sci. Rev*. 20, 451

Conrath, B. J. 1969 On the estimation of relative humidity profiles from medium resolution infrared spectra obtained from a satellite. *J. Geophys. Res*. 74, 3347

Conrath, B. J., Hanel, R. A., Kunde, V. G., and Prabhakara, C. 1970 The infrared interferometer experiment on Nimbus 3. *J. Geophys. Res*. 75, 5831-5857

Conrath, B. J. 1972 Vertical resolution of temperature profiles obtained from remote radiation measurements, *J. Atmos. Sci*. 29, 1262-1271

Conrath, B. *et al*. 1973 Atmospheric and surface properties of Mars obtained by infrared spectroscopy on Mariner 9. *J. Geophys. Res*. 78, 4267-4278

Conrath, B. I. 1981 Planetary Scale wave structure in the Martian Atmosphere. Icarus 48, 246-255

Coy, L. and Leovy, C. 1977 A comparison of midwinter stratospheric warmings in the southern and northern hemispheres. *Proc. IAMAP/IAGA Assembly*, Seattle

Crane, A. J. 1979a Aspects of the energetics of the upper stratosphere during the January-February 1973 major sudden warming. *Quart. J. R. Met. Soc*. 105, 231-252

Crane, A. J. 1979b Annual and semiannual waves in the temperature of the mesosphere as deducted from Nimbus 6 PMR measurements. *Quart. J.R. Met. Soc*. 105, 509-520

Crane, A., Haigh, J. D., Pyle, J. A., and Rogers, C. F. 1980 Mean meridional circulations of the stratosphere and mesosphere. *Pageoph*. 118, 307-32

Cressman, G. P. 1959 An operational objective analysis system. *Mon. Weath. Rev*. 85, 367-374

Cross, M. J. 1972 Nimbus 4 selective chopper radiometer data handling. *Computer J*. 172-180

Crutzen, P. J. 1970 *Q.J.R. Met. Soc*. 96, 769-70

Curtis, P. D. *et al*. 1973 A pressure modulator radiometer for high altitude temperature sounding. *Proc. R. Soc. Lond*. A336. 1

Curtis, P. D., Houghton, J. T., Peskett, G. D. and Rodgers, C. D. 1974 Remote sounding of atmospheric temperature from satellites V. The pressure modulator radiometer for Nimbus F. *Proc. R. Soc. Lond*. A337, 135-150

Curtis, P. D. and Houghton, J. 1975 Un radiometer satellitaire pour determinerles temperatures de la stratosphere. *La Meteorologie*, Societe Meteorologique de France, Nov. 1975, 105-118

Dave, J. V. and Furukawa P. M. 1967 *J. Atmos. Sci* 24, 175-81

Dave, J. V. and Mateer E. L. 1967 A preliminary study on the possibility of estimating total atmosphere ozone from satellite measurements. *J. Atmos. Sci*. 24, 414-427

Davis, P. A., Evans, W. F., Mancuso, R. L. and Wold, D. E. 1976 Height positioning on cloud motions from infra-red tracking. *Proceedings of the symposium on meteorological observations from space and their contribution to the FGGE. COSPAR.* 211-213

Deland, R. J. 1977 Evidence of downward propagating planetary-scale waves in the southern mesosphere winter stratosphere. *Proc. IAMAP/IAGA Assembly*, Seattle

DeLuisi, F. J. and Mateer, C. C. 1971 On the application of the optimum statistical inversion technique to the evaluation of Umkehr observation. *J. Appl. Meteorol.* 10, 328

Desmarais, A., Tracton, M. S., McPherson, R. S. and Van Haaren, R. 1978 *The NMC report of the data systems test (NASA contract S-70252-AG).* US Dept. of Commerce National Oceanic and Atmospheric Administration, Washington, D.C.

Dickinson, R. E. 1969 Vertical propagation of planetary waves through an atmosphere with Newtonian cooling. *J. Geophys. Res.* 74, 929-938

Dickinson, R. E. 1975 Energetics of the stratosphere. *J. Atmos. Terr. Phys.* 37, 855-864

Drummond J. R. Houghton J. T., Peskett G. D., Rodgers C. D., Wale M. J., Whitney J, Williamson E. J. 1980 The stratospheric and mesospheric sounder on Nimbus 7. *Phil Trans R. Soc London*, A296, 219-241

Drummond J. R. and Mutlow C. T. 1981 Satellite measurement of H₂O fluorescence in the mesosphere. *Nature* 294, 431-432

Eccles, D. and King, J. W. 1969 Proc. IEEE 57, 1012-88

Elliot, D. D., Clar, M. A. and Hudson, R. D. 1967 *Aerospace techn. report No. TR-0158 (3260-10).* 2 September 1967

Elliot, D. 1971 Effects of a high altitude (50 km) aerosol layer on topside ozone sounding. *Space research XI*, 857-861 Akademie-Verlag

Ellis, P. J. et al. 1970 First results from the selective chopper radiometer on Nimbus 4. *Nature* 228, 139-143

Ellis, P. J. et al. 1972 Infra-red atmospheric temperature sounding from satellites. *Radio and electronic engineer*, 42, 155-161

Ellis, J. S., Vonder Haar, T. H., Levitus, S. and Oort, A. H. 1978 The annual variation in the global heat balance of the earth. *J. Geophys. Res.* 83, 1958-1962

Elson, L. S. 1979 Preliminary results from the Pioneer Venus orbiter infrared radiometer: temperature and dynamics in the upper atmosphere. *Geophys. Res. Lett.* 8, 720-722

Ensor, L. J. 1978 *User's guide to the operation of the NOAA Geostationary satellite system.* US dept. of Commerce, NOAA, Washington, D.C.

Esposito, L. W. et al. 1979 Sulphur dioxide in the Venus atmosphere. *Geophys. Res. Lett.* 6, 601-604

Eyre, J. R. 1981 Meteosat water-vapour imagery. *Met Mag.* 110, 345-351

Fabian, P., Pyle, J. A. and Wells, R. J. 1979 The August 1972 solar proton event and the atmospheric ozone layer. *Nature*, 277, 458-460

Farmer, C. B. and Houghton, J. T. 1966 Collision-induced absorption in the Earth's atmosphere. *Nature* 209, 1341

Farmer, C. B. and Laporte, D. D. 1972 The detection and mapping of water vapour in the Martian atmosphere. *Icarus* 10, 34

Farmer, C. B. et al. 1977 Mars: water vapour observations from the Viking orbiter. *J. Geophys. Res.* 28, 4225-4248

Fjeldbo, G. et al. 1971 The neutral atmosphere of Venus as studied with the Mariner 5 Dual-frequency occultation experiment. *Astron. J.* 76, 123-140

Fjeldbo, G., Seidel, B. and Sweetnam, D. 1975 The Mariner 10 radio occultation measurements of the ionosphere of Venus. *J. Atmos. Sci.* 32, 1232-1236

Flattery, T. 1970 Spectral models for global analysis and forecasting. Proceedings sixth

AWS technical exchange conference, U.S. Naval Academy. *Air Weather Service Technical Report* **242**, 42-54. (pub. 1971)

Frank, L. A. *et al.* 1981 Global auroral imaging instrumentation for the Dynamics Explorer mission. *Space Science Instrumentation* **5**, 369-393

Franklin, J. N., 1970 Well-posed stochastic extensions to ill-posed linear problems. *J. Math. Anal. Appl.* **31**, 682

Fraser, G. T. 1976 The covariance of temperature and ozone due to planetary-wave forcing. *NASA preprint X-991-76-192*

Frederick, J. E., Hays, P. B., Guenther, B. W. and Heath, D. F. 1977*a* Ozone abundances in the lower mesosphere deduced from backscattered solar radiances, *J. Atmos. Sci.* **34**, 1987-1999

Frederick, J. E., Guenther, B. W., and Heath, D. F. 1977*b* Spatial variations in tropical ozone: the influences of meridional transport and planetary waves in the stratosphere, *Beitr. Phys. Atmosphere* **50**, 496-507

Frederick J. E., Guenther, B., Hays, P. B. and Heath, D. F. 1978 Ozone profiles and chemical loss rates in the tropical stratoshpere deduced from backscattered ultraviolet measurements, *J. Geophys. Res.* **83**, 953-958

Fritz, S. and Rao, P. K. 1967, *J. Appl. Met.* **6**, 1088-96

Fritz, S. and Soules, S. D. 1972 Planetary variations of stratosphere temperatures. *Mon. Weath. Rev.* **100**, 582-589

Fujita, T. T., Watanabe, K. and Izawa, T. 1969 Formation and structure of equatorial anticyclones caused by large scale cross-equatorial flows determined by *ATS-1*, *Mon. Weath. Rev.* **8**, 649-667

Fujita, T. T., Pearl E. W. and Shenk, W. E. 1975 Satellite-tracked cumulus velocities *J. Appl Met.* **14**, 407-413

Gaby, D. C. and Poteat, K. O. 1973 ATS-3 satellite-derived low level winds: a provisional climatology, *J. Appl. Met.* **12**, 1054-1061

Gandin, L. S. 1963 *Objective analysis of meteorological fields.* Gidrometeorologicheskoe Izdatel'stvo (DIMIZ). Leningrad. English translation by Israel program for scientific translations, Jerusalem, 1965 (available from NTIS)

Garcia, O. 1981 A comparison of two satellite rainfall estimates for GATE. *J. Appli. Met.* **20**, 430-438

Geisler, J. E. and Dickinson, R. E. 1976 The five-day wave on a sphere with realistic zonal winds. *J. Atmos. Sci.* **33**, 632-641

Georgii, H. W. and Jost, D. 1969 *Nature* **221**, 1040

Ghazi, A. 1977 Stratospheric thermal wave - ozone field interactions. *Proc. IAMAP/IAGA Assembly*, Seattle

Gilchrist, A. 1982 J. S. C. Study Conference on Observing Systems Experiments, GARP and WCRP Numerical Experimentation Programme Report No. 4, W.M.O. Geneva

Gille, J. C. and House, F. B. 1971 On the inversion of limb radiance measurements 1: temperature and thickness. *J. Atmos. Sci.* **29**, 1427-1442

Gille, J. C. 1972 *Temperature, its measurement and control in science and industry*, vol 4 (New York: American Institute of Physics)

Gille, J. C., Bailey, P., House, F. B., Craig, R. A. and Thomas J. R. 1975 in *Nimbus 6 User's Guide* (ed. J. E. Sissala) Greenbelt, Md.: National Aeronautics and Space Administration

Gille, J. C. *et al.* 1978 The limb infra-red monitor of the stratosphere experiment: *Nimbus 7 User's Guide*. NASA (Goddard Space Flight Centre)

Gille, J. C. and Bailey, P. L. 1978 In *Remote Sensing of the Atmosphere: Inversion Methods and Applications* (ed. A. L. Fymat and V. E. Zuev) 101-113. Amsterdam: Elsevier

Gille, J. C., Bailey, P. L. and Russell, J. M. III 1980. Temperature and composition measurements from the LRIR and LIMS experiments on Nimbus 6 and 7. *Phil Trans R. Soc. Lond* A296, 205-218

Gloersen, P., Wilheit, T. T., Chang, T. C., Nordberg, W. and Campbell, W. J. 1974 Microwave maps of the polar ice of the earth. *Bull. Amer. Met. Soc.* 55, 1442-1448

Gloersen, P. and Hardis, L. 1978 The Scanning Multichannel Microwave Radiometer experiment (SMMR). *Nimbus 7 User's Guide.* NASA (Goddard Space Flight Centre)

Gnedykh *et al.* 1976 Preliminary results of studying the infrared spectrum of Venus from Venera 9 and 10. *Kosicheskiye Issledovaniya* 14, 758-767

Golovko, V. A. and Spankich, D. 1978 In *Remote sensing of the atmosphere: inversion methods and application* (ed. A. L. Fymat and W. E. Zuev), Amsterdam: Elsevier. 91-95

Gonzalez, F. I. *et al.* 1979 Seasat synthetic aperture radar - ocean wave detection capabilities. *Science* 204, 1418-1421

Goody, R. M. 1964 *Atmospheric Radiation.* Oxford University Press.

Gringuaz, K. I. and Breus, T. K. 1970 Comparative characteristics of the ionospheres of the planets of the terrestrial group. *Sp. Sci. Rev.* 10, 743-769

Grody, N. C., Gruber, A. and Shen, W.C. 1980 Atmospheric water content over the tropical Pacific derived from the Nimbus 6 Scanning Microwave spectrometer. *J Appl Met.* 19, 986-996

Groves, G. V. 1970 *Seasonal and latitudinal models of atmospheric temperature, pressure and density, 25 to 100 km,* AFCRL-70-0261

Guenther, R., Dasgupta, R. and Heath, D. 1977 Twilight ozone measurements by solar occultation from AE-5, *Geophys. Res. Lett.* 4, 434-437

Guenther, B. and Dasgupta, R. 1979 High resolution measurements of upper level column content in the tropical ozone fields. Manuscript in preparation

Gurvich, A. S. and Demin, V. V. 1970 *Isv. Atmos. Ocean. Phys.* 6, 771-779

Guymer, L. B. 1978 *Operational applications of satellite imagery to synoptic analysis in the southern hemisphere.* Technical report No 29, Department of Science, Australian Bureau of Meteorology, Melbourne

Halem, M., Ghil, M., Atlas, R., Susskind, J. and Quirk, W. 1978 *The GISS temperature sounding impact test,* NASA Technical Memorandum 78063

Hanel, R. A. and Conrath, B. J. 1969 Interferometer experiment on Nimbus 3: preliminary results, *Science* 165, 1258-1260

Hanel, R. A. 1970 Recent advances in satellite radiation measurements *Adv. Geophys.* 14, 359 (New York: Academic Press)

Hanel, R. A., Schlachman, B., Clark, F. D., Prokesh, C. H., Taylor, J. B., Wilson, W. M. and Chaney, L 1970 The Nimbus 3 Michelson interferometer. *Applied Optics.* 9, 1967-1970

Hanel, R. A., Schlachman, B., Rogers, D. and Vanous, D. 1971 Nimbus-4 Michelson Interferometer. *Applied Optics.* 10, 1376-1382

Hanel, R. *et al.* 1972 Investigation of the Martian environment by infrared spectroscopy on Mariner 9. *Icarus* 17, 423-442

Hanel, R. *et al.* 1979 Infrared observations of the Jovian system from Voyager 1. *Science* 204, 972-976

Hanel, R. *et al.* 1980 Infrared spectrometer for Voyager. *Applied Optics* 19, 1391-1400

Hanel, R. *et al.* 1981 Infrared observations of the Saturnian system from Voyager 1. *Science* 212, 192-200

Hansen, J. E. and Hovenier, J. W. 1974 Interpretation of the polarization of Venus. *J. Atmos. Sci.* 31, 1137-1160

Harries, J. E., Llewellyn-Jones, D., Saunders, R. and Zavody, A. M. 1983 Observation of sea surface temperature for climate research. *Phil Trans R. Soc.* (1983) - to be published

Hartmann, D. L. 1976 The structure of the stratosphere in the southern hemisphere during late winter 1973 as observed by satellite. *J. Atmos. Sci.* 33, 1141-1154

Hartmann, D. L. 1976 The dynamical climatology of the stratosphere in the southern hemisphere during late winter 1973. *J. Atmos. Sci.* 33, 1789-1802

Hartmann, D. L. 1976 Dynamic studies of the southern hemisphere stratosphere. *COSPAR Space Research XVII*, 167-174

Hartmann, D. L. 1977 Comments on stratospheric long waves: comparison of thermal structure in the northern and southern hemisphere. *J. Atmos. Sci.* 34, 434-436

Hartmann, D. L. 1977 On potential vorticity and transport in the stratosphere. *J. Atmos. Sci.* 34, 968-977

Hartmann, D. L. and Garcia, R. R. 1979 *J. Atmos. Sci.* 36, 1

Harwood, R. S. and Pyle, J. A. 1974 A time; dependent two dimensional model of the atmosphere below 80 km. Proceedings of the International Conference on structure, composition and general circulation of the upper and lower atmosphere and possible anthropogenic perturbations. Melbourne, January, 1974.

Harwood, R. S. 1975 The temperature structure of the southern hemisphere stratosphere August-October 1971. *Quart. J. R. Met. Soc.* 101, 75-91

Harwood, R. S. and Pyle, J. A. 1975 A two-dimensional mean circulation model for the atmosphere below 80 km. *Quart. J. R. Met. Soc.* 101, 723-747

Harwood, R. S. 1976 Some recent investigations of the upper atmosphere by remote sounding satellites. *Colston Papers* 28, (Ed. Curtis and Barrett). Bristol.

Harwood, R. S. and Pyle, J. A. 1977 Studies of the ozone budget using a zonal mean circulation model and linearized photochemistry. *Quart. J. R. Met. Soc.* 103, 319-343

Hasler, A. F. 1981 Stereographic Observations from geosynchronous satellites: an important new tool for the atmospheric sciences. *Bull Amer. Met. Soc.,* 62, 194-212

Hasler, A. F., Shenk, W. E. and Skillman, W. C. 1977 Wind estimates from cloud motions: results from phases I, II and III of an *in situ* aircraft verification experiment, *J. Appl. Met.* 16, 812-815

Hawson, C. L. 1970 Performance requirements of aerological instruments: *WMO Technical Note 112*, WMO-No. 267. TP 151

Hayden, C. M. 1971 Nimbus 3 SIRS pressure height profiles as compared to radiosondes *Mon. Weath, Rev.* 99, 659-64

Hayden, C. M., Hubert, L. F., McClain, E.P., Seaman, R. S. 1979 Quantitative meteorological data from satellites. *WMO Technical Note 166,* World Meteorological Organization, Geneva

Hayden, C. M., Smith, W. L. and Woolf, H. M. (1981) Determination of moisture from NOAA Polar Orbiting Satellite Sounding radiances. *J. Appl. Met.* 20, 450-466

Hays, P. B., Killeen, T. C. and Kennedy, B. C. (1981) The Fabry-Perot interferometer on Dynamics Explorer. *Space Science Instrumentation,* 5, 395-416

Hays, P. B. and Olivero J. J. 1970 *Planet. Space Sci.* 18, 1729-33

Hays, P. B. and Roble R. G. 1973 Observation of mesospheric ozone at low latitudes, *Planet. Space Sci.* 21, 273-279

Heasman, C. C. and Crane, A. J. 1978 Proceedings of COSPAR symposium, Innsbruck. In press.

Heath, D. and Westcott, R. D. 1970 *Nimbus-4 User's Guide* 135-48 NASA (Goddard Space Flight Centre).

Heath, D. 1973 Space observations of the variability of solar irradiance in the near and far ultraviolet, *J. Geophys. Res.* 78, 2779-2792

Heath, D. F., Mateer, C. L. and Krueger, A. J. 1973 The Nimbus-4 backscatter ultraviolet (BUV) atmosphere ozone experiment-two year's operation, *Pure and Appl. Geophys.* 106-108, 1239-1253

Heath, D. F., Krueger, A. J., Roeder, H. A. and Henderson, B. D. 1975 The solar backscatter ultraviolet and total ozone mapping spectrometer (SBUV/TOMS) for Nimbus G, *Opt. Engr.* 14, 323-331

Heath, D. F., Krueger, A. J. and Crutzen, P. J. 1977 Solar proton event influence on stratospheric ozone. *Science* 197, 323-331

Herman, B. M. and Yarger, D. N. 1969 Estimating the vertical atmospheric ozone distribution by inverting the radiative transfer equation for pure molecular scattering. *J. Atmos. Sci.* 26, 133

Herr, K. C. *et al.* 1972 Mariner Mars 1969 infrared spectrometer. *Applied Optics,* 11, 493-501

Hickey, J. R., Alton, B. M., Griffin, F. J., Jacobowitz, H., Pellegrino, P. and Smith, E. A. 1982. Observations of the solar constant and its variations with emphasis on Nimbus 7 results. *The symposium on the solar constant and the spectral distribution of solar irradiance,* IAMAP, Hamburg, 1981. P. 10-17. Extended abstracts published by the Radiation Commission, 1982.

Hillger, D. W. and Vonder Haar, T. M. 1977 Deriving mesoscale temperature and moisture fields from satellite radiance measurements over the United States. *J. Appl. Met.* 16, 715-726

Hillger, D. W. and Vonder Haar, T. H. 1981 Retrieval and use of high resolution moisture and stability fields from Nimbus 6 HIRS radiances in pre-convective situations. *Mon. Wea Rev.,* 109, 1788-1806

Hilsenrath, E., Heath, D. B., Schlesinger, M. 1978 *Seasonal and interannual variations in total ozone revealed by the Nimbus-4 backscattered ultraviolet experiment.* NASA Technical Memorandum 79695. Dec. 1978

Hirota, I. 1971 Excitation of planetary Rossby waves in the winter stratosphere by periodic forcing. *J. Met. Soc. Japan* 49, 439-448

Hirota, I. *et al.* 1973 Structure and behaviour of the Aleutian anticyclone as revealed by meteorological rocket and satellite observations. *J. Met. Soc. Japan* 51, 353-363

Hirota, I. 1975 Spectral analysis of planetary waves in the summer stratosphere and mesosphere. *J. Met. Soc. Japan* 53, 33-34

Hirota, I. 1976 Seasonal variation of planetary waves in the stratosphere observed by the Nimbus 5 selective chopper radiometer. *Quart. J. R. Met. Soc.* 102, 757-770

Hirota, I. and Barnett, J. J. 1977 Planetary waves in the winter mesosphere - preliminary analysis of Nimbus-6 PMR results. *Quart. J. R. Met. Soc.* 103, 487-498

Hirota, I. 1978 Equatorial waves in the upper stratosphere and mesosphere in relation to the semiannual oscillation of the zonal wind. *J. Atmos. Sci.* 35, 714-722

Hirota, I. 1979 Kelvin waves in the equatorial middle atmosphere observed by the Nimbus-5 SCR. *J. Atmos. Sci.* 36, 217-222

Hollinger, J. P. 1971 Remote passive microwave measurements of the sea surface. AGARD Conference Proceedings No. 90 on Propagation Limitations in Remote Sensing AGARD-CP-90-71, 14.1 - 14.7

Holton, J. 1975 The dynamic meteorology of the stratosphere and mesosphere. *Met. Mon.* 15, Amer. Met. Soc.

Holton, J. and Mass, C. 1976 Stratospheric vacillation cycles, *J. Atmos. Sci.* 33, 2218-2225

Holton, J. R. 1976 A semi-spectral numerical model for wave mean flow interactions in the stratosphere: Applications to sudden stratospheric warmings. *J. Atmos. Sci.* 33, 1639-1649

Iord, C. W. *et al.* 1970 Ultraviolet spectroscopy experiment for Mariner Mars. 1971. *Icarus* 12, 63-77

Horn, D. *et al.* 1972 The composition of the Martian atmosphere: Minor constituents. *Icarus* 16, 543

Houghton, J. T. 1961 The meteorological significance of remote measurements of infrared emission from atmospheric carbon dioxide. *Quart. J. R. Met. Soc.* 87, 102-104

Houghton, J. T. 1963 Stratospheric temperature measurements from satellites, *J. Brit. Interplan. Soc.* 19, 382-386

Houghton, J. T. and Smith, S. D. 1966 *Infra-red Physics* (London: Oxford University Press)

Houghton, J. T. 1969 Absorption and emission by carbon dioxide in the mesosphere. *Q. J. R. Met. Soc.* 95, 1-20

Houghton, J. T. and Smith, S. D. 1970 Remote sounding of atmospheric temperature from satellites. *Proc. R. Soc. Lond.* A320, 23-33

Houghton, J. T. and Hunt, G. E. 1971 The detection of ice clouds from remote measurements of their emission in the far infra-red. *Quart. J. R. Met. Soc.* 97, 1-17

Houghton, J. T. 1971 The selective chopper radiometer on Nimbus 4. *Bull Amer. Met. Soc.* 53, 27-28

Houghton, J. T. 1972 The selective chopper radiometer on Nimbus 4. *Bull. Amer. Met. Soc.* 53, 27-28

Houghton, J. T. and Lee, A. C. L. 1972 Atmospheric transmission in the 10-12 μm window, *Nature* 238, 117-118

Houghton, J. T. and Taylor, F. W. 1973 Remote sounding from artificial satellites and space probes of the atmosphere of the Earth and the planets. *Rep. Prog. Phys.* 36, 827-919

Houghton, J. T. 1977 Calibration of infra-red instruments for the remote sounding of atmospheric temperature. *Applied Optics* 16, 319-321

Houghton, J. T. 1978 The stratosphere and mesosphere. *Quart. J. R. Met. Soc.* 104, 1-29

Houghton, J. T. 1979 The future role of observations from meteorological satellites. *Quart. J. R. Met. Soc.* 105, 1-28

Houghton, J. T. 1981 Remote sounding of the atmosphere and ocean for climate research. *Proc. IEE* 128, 442-448

Houghton, J. T. (ed) 1983 The Global Climate, Cambridge University Press

Hubert, L. F. 1976 Wind determination from geostationary satellites. *Proceedings of the Symposium on Meteorological Observations from Space: their contribution to the FGGE* COSPAR 211-213

Ingersoll, A. P. *et al.* 1976 Results of the infrared radiometer experiment on Pioneers 10 and 11. (In *Jupiter*, ed. T. Gehrels). U. Arizona Press, Tuscon, 197-205

Iozenas, V. W., Krasnopol'sky, A., Kuznetsov, and Lebedinskiy, 1969 Studies of the earth's zonosphere from satellites. *Atmos and Oceanic Phys.* 5, 149-159

Jacobowitz, H., Smith, W. L., Howell, H. B. and Nagle, F. W. 1979 The first 6 months of radiation budget measurements from the Nimbus 6 ERB experiment. *J. Atmos. Sci.* 36, 501-507

Jacquinot, P. 1954 *J. Opt. Soc. Am.* 44, 761-5

Johnson, K. W. and Rodenhuis, D. R. 1977 Potential vorticity transport during a stratospheric warming. *Proc. IAMAP/IAGA Assembly*, Seattle

Johnson, K. W. 1977 Variations in static stability in the stratosphere and lower mesosphere during a winter disturbance. *Proc. IAMAP/IAGA Assembly*, Seattle

Jones, T. S. and Williamson, E. J. 1973 *The analysis of data from meteorological satellites. Les Satellites Météorologics, C.N.E.S.* Paris, 351-362

Jones, W. L. *et al.* 1979 Seasat scatterometer: Results of the Gulf of Alaska Workshop. *Science* 204, 1413-1415

Kaplan, L. D. 1959 Inference of atmospheric structure from remote radiation measurement. *J. Opt. Soc. Amer.* **49**, 1004

Kastner, M., Fischer, H. and Bolle, H.-J. 1980 Wind determination from Nimbus 5 observations in the 6.3 μm water vapour band. *J. Appl. Met.* **19**, 409-418

Kawabata, K. *et al.* 1980 Cloud and haze properties from Pioneer Venus photopolarimetry. *J. Geophys. Res.* **85**, 8129-8140

Kelly, G. A. M. 1977 A cycling experiment in the southern hemisphere using VTPR data. Australian Numerical Meteorology Research Centre, Melbourne. Unpublished manuscript

Kelly, G. A. M., Mills, G. A. and Smith, W. L. 1978 Impact of Nimbus-6 temperature soundings on Australian Region Forecasts. *Bull. Amer. Met. Soc.* **59**, 393-405

Kennedy, J. S. and Nordberg, W. 1967 *J. Atmos. Sci.* **24**, 711-719

Kidder, S. Q. and Vonder Haar, T. H. 1976 A comparison of satellite rainfall estimation techniques over the GATE area. *Proceedings of the symposium on meteorological observations from space.* 123-125, Philadelphia, COSPAR

Kidder, S. Q. and Vonder Haar, T. H. 1977 Seasonal oceanic precipitation frequencies from Nimbus 5 microwave data. *J. Geophys. Res.* **82**, 2083-2086

Kieffer, H. H. *et al.* 1972 Infrared thermal mapping experiment. The Viking Mars Orbiter. *Icarus* **16**, 47-56

Kieffer, H. H. *et al.* 1977 Thermal and albedo mapping of Mars during the Viking primary mission. *J. Geophys. Res.* **82**, 4249-4291

Killeen, T. L., Hays, P. B., Spencer, N. W. and Wharton, L. F. 1982 Neutral winds in the polar thermosphere as measured from Dynamics Explorer. *Geophys. Res. Lett.* **9**, 959-960

King, J. I. F., 1966 Inversion by slabs of varying thickness. *J. Atmos. Sci.* **21**, 324

Kliore, A. 1968 In *Radio occultation measurements of the atmosphere of Mars and Venus* (ed. Brandt and McElroy) 205-224. Gordon and Breach: New York

Kliore, A. *et al.* 1972 The atmosphere of Mars from Mariner 9 radio occultation experiments. *Icarus,* **17**, 484-516

Kliore, A and Patel, I. R. 1980 Vertical structure of the atmosphere of Venus from Pioneer Venus radio occultation. *I. Geophys. Res.* **85**, 7957-7962

Kondrat'yev, K. Ya. *et al.* 1970 *Izv. Acad. Sci USSR Atmos. Ocean Phys.* **6**, 388-411

Kransnopolsky, V. A. *et al.* 1976 Spectroscopy of the night-sky luminescence of Venus from Venera 9 and 10. *Kosnicheskic Issledovaniya* **14**, 789-795

Krueger, A. J., Heath, D. F. and Mateer, C. L. 1973 Variations in the stratospheric ozone field inferred from Nimbus satellite observations. *Pure and Appl. Geophys,* **106-108**, 1254-1263

Krueger, A. J., Guenther, B., Fleig, A. J., Heath, D. F., Hilsenrath, E., McPeters, R. and Prabhakara, C. 1980 Satellite ozone measurements. *Phil. Trans R. Soc. London,* A**296**, 191-204

Krueger, A. J. and Minzer, R. A. 1976 A mid-latitude ozone model for the 1976 U.S. Standard Atmosphere. *J. Geophys. Res.* **81**, 4477-4481

Ksanfomality, L. V. 1980 Venera 9 and 10: Thermal Radiometry. *Icarus,* **41**, 36-64

Kumar, S. 1976 Mercury's atmosphere: A perspective after Mariner 10. *Icarus* **28**, 579-591

Kurt, V. G. *et al.* 1976 *Hydrogen observations at Lyman alpha of the upper atmosphere of Venus with Venera 9 and 10.* Presented at COSPAR, 1976

Labitzke, K. 1971 *Synoptic-scale motions above the stratopause.* NCAR ms. 71-139. Presented at IUGG Assembly, Moscow

Labitzke, K. 1972 Temperature changes in the mesosphere and stratosphere connected with circulation changes in winter. *J. Atmos. Sci.* **29**, 756-766

Labitzke, K. 1972 The interaction between stratosphere and mesosphere in winter. *J. Atmos. Sci.* 29, 1395-1399

Labitzke, K. and Barnett, J. J. 1973 Global time and space changes of satellite radiances received from the stratosphere and low mesosphere. *J. Geophys. Res.* 78, 483- 496

Labitzke, K. 1974 The temperature in the upper stratosphere: differences between hemispheres. *J. Geophys. Res.* 79, 2171-2175

Labitzke, K. 1976 Comparison of the stratospheric temperature distribution over northern and southern hemispheres. *COSPAR Space Research XVII*, 159-163

Labitzke, K. 1976 On the use of single channel radiances for estimating temperatures at discrete pressure levels in the upper stratosphere. *COSPAR Space Research XVII*, 151-157

Labitzke, K. and Barnett, J. J. 1979 Review of climatological information obtained from remote sensing of the stratosphere and mesosphere. *COSPAR Space Res.* XIX, 97-106

Lane, A. L. *et al.* 1973 Mariner 9 UV spectrometer observations of ozone on Mars. *Icarus* 18, 102

Lawrence, G. M., Barth, C. A., and Argabright, V. 1977 Excitation of the Venus night airglow. *Science* 195, 573

Lenoir, W. B. 1968 *J. Geophys. Res.* 73, 361-376

Leovy, C. *et al.* 1972 The Martian atmosphere: Mariner 9 television experiment progress report. *Icarus*, 17, 373

Leovy, C. B. and Webster, P. J. 1976 Stratospheric long waves: comparison of thermal structure in the northern and southern hemispheres. *J. Atmos. Sc.* 33, 1624-1638

Lienesch, J. H. and Wark, D. Q. 1967 *J. App. Met.* 6, 674-82

Liljas, E. 1982 Automated Techniques for the analysis of satellite cloud imagery. In *Nowcasting* ed. Browning, Academic Press, pp 167-176

Lipes, R. G. *et al.* 1979 Seasat Scanning Multichannel Microwave Radiometer: Results of Gulf of Alaska Workshop. *Science* 204, 1415-1417

London, J., Bojkov, R. D., Oltmans and Kelley, J. I. 1968 *Atlas of the Global Distribution of total ozone, July 1957-June 1967,* NCAR Tech. Note TN/113+STR, National Centre for Atmospheric Research, Boulder, Colorado.

London, J., Frederick, J. E. and Anderson, G. P. 1977 Satellite Observation of the global distribution of stratospheric ozone. *J. Geophys. Res.* 82, 2543-2556

Lorenc, A., Rutherford, I and Larsen, G. 1977 *The ECMWF analysis and data-assimilation scheme - analysis of mass and wind fields.* ECMWF Technical Report No. 6, 46

Lovejoy, S. and Austin, G. L. 1979 *Mon. Weath. Rev.* 107, 1048-1051

Lovill, J. E. *et al.* 1978 *Total ozone retrieval from satellite multichannel filter radiometer measurements.* Lawrence Livermore Laboratory UCRL-53473, May 25

Machenhauer, B. 1977 On the dynamics of gravity oscillations in a shallow-water model, with applications to normal mode initialization. *Contrib. Atmos. Phys.* 50, 253-271

Madden, R. A. and Julian, P. A. 1973: Reply to comments by R. J. Deland on further evidence of global scale, 5-day pressure waves. *J. Atmos. Sci.* 30, 935-940

Maddox, R. A. and Vonder Haar, T. H. 1979 Covariance analysis of satellite-derived mesoscale wind fields, *J. Appl. Met.* 18. 1327-1334

Mariner Stanford Group 1979 Venus-Ionosphere and atmosphere as measured by dual-frequency radio occultation of Mariner 5. *Science* 158, 1678-1683

Mateer, C. L. 1965 *J. Atmos. Sci.* 22, 370-81

Mateer, C. L., Heath, D. F. and Krueger, A. J. 1971 Estimation of total ozone from satellite measurements of backscattered ultraviolet earth radiance. *J. Atmos. Sci.* **28**, 1307-1311

Mateer, C. L. 1972 A review of some aspects of inferring of ozone profile by inversion of ultraviolet radiance measurements. In *Mathematics of Profile Inversion*, (ed. Colin). NASA TMX-62, 150, 2-25

Matsuno, M. 1970 Vertical propagation of stationary planetary waves in the winter northern hemisphere. *J. Atmos. Sci.* **27**, 871-883

Matsuno, M. 1971 A dynamical model of the stratospheric sudden warming *J. Atmos. Sci.* **28**, 1479-1494.

McCleese, D. J. and Wilson, L. S. 1976 Cloud top heights from temperature sounding instruments *Quart, J. R. Met. Soc.* **102**, 781-790

McCleese, D. J. 1978 Remote sensing of cloud properties from Nimbus 5. Reprinted from *Remote sensing of the Atmosphere: Inversion methods and applications*, (ed. Fymat and Zuer), 295-304. Elsevier.

McCleese, D. J. 1978 Remote sensing of cloud properties from Nimbus 5. *Remote sensing of the Atmosphere: Inversion methods and applications*, (ed. Fymat and Zuev), 295-304. Elsevier, Amsterdam.

McCormick, M. P. *et al.* 1979 Satellite studies of the stratospheric aerosol. *Bull. Am. Met. Soc.* **60**, 1038-1046

McCormick, M. P. *et al.* 1981 High latitude stratospheric aerosols measured by the SAM II satellite system in 1978 and 1979. *Science* **214**, 328-331

McCormick, M. P. 1982 Global Distribution of stratospheric aerosols by satellite measurements. AIAA 20th Aerospace Sciences Meeting, Orlando, Florida 1982

McCormick, M. P., Steele, H. M., Hamill, P., Chu, W. P., Swissler, T. J. 1982 Polar stratospheric cloud sightings by SAM II *J. Atmos Sci*, **39**, 1387-1397

McGregor, J. and Chapman, W. A. 1978 Observations of the annual and semi-annual wave in the stratosphere using Nimbus 5, SCR data. *J. Atmos. Terr. Phys.* **40**, 677-686

McPherson, R and Kistler, R. 1974 *Real-data assimilation experiments with implicit and explicit integration methods.* Office Note 96, NOAA, National Weather Service, National Meteorological Centre.

Meeks, M.L. and Lilley, A.E. 1963 *J.Geophys. Res.* **68**, 1683-1703.

Miller, D. E. and Stewart, K. H. 1968 Observations of atmospheric ozone from an artificial earth satellite. *Proc. R. Soc.* **A288**, 540-544

Miller, D. E., Brownscombe, J. L., Carruthers, G. P., Pick, D. R. and Stewart, K. H. 1980 Operational temperature sounding of the stratosphere. *Phil. Trans. R. Soc. Lond.* **A292**, 65-71

Miller, A. J. and Hayden, C. M. 1978 The impact of satellite-derived temperature profiles on the energies of NMC analyses and forecasts during the August 1975 Data Systems Test, *Mon. Weath. Rev.* **106**, 390-398

Mills, G. A. and Smith, W. L. 1978 Impact of Nimbus-6 temperature soundings on Australian region forecasts. *Bull. Am. Soc.* **59**, 393-405

Molnar, Gy. 1979 Interpretation of Nimbus 5 SCR data obtaining vertical temperature profiles. *Quart. J. R. Met. Soc.* **105**, 461-467

Moroz, V. I., and Ksanfomality, L. V. 1972 Preliminary results of astrophysical observations of Mars from Mars-3. *Icarus* **17**, 408

Muller, H. G. and Nelson, L. 1978 A travelling quasi 2-day wave in the meteor region. *J. Atmos. Terr. Phys.* **40**, 761-766, 1978

Murray, B. C. *et al.* 1974 Venus atmospheric motion and structure from Mariner 10 pictures *Science* **183**, 1037-1315

imbus Project, 1978 The Nimbus 7 User's Guide. NASA (Goddard Space Flight Centre).

Ohring, G. 1979 Impact of satellite temperature sounding data on weather forecasts. *Bull. Amer. Met. Soc.* 60, No. 10, 1142-1147

Oort, A. H. and Vonder Haar, T. H. 1976 On the observed annual cycles in the ocean-atmosphere heat balance over the northern hemisphere. *J. Phys. Oceanog.* 6, 781-800

Orton, G. S. *et al.* 1975 The thermal structure of Jupiter. *Icarus*, 26, 125-158

Park, S.-U., Sikdar, D. N. and Suomi, V. E. 1974 Correlation between cloud thickness and brightness using Nimbus 4 THIR data and ATS-3 digital data. *J. Appl. Met.* 13, 402-410

Pearce, J. B. *et al.* 1971 The Mariner 6 and 7 UV spectrometer. *Applied Optics* 10, 805-812

Peckham, G., Rodgers, C. D., Houghton, J. T. and Smith, S. D. 1967 *In Electro-magnetic Sensing of the Earth from satellites*, (ed Zirkind). New York: Press of the Polytechnic Institute of Brooklyn

Peckham, G. 1974 The information content of remote measurements of atmospheric temperature by satellite infra-red radiometry and optimum radiometer configuration. *Quart. J. R. Met. Soc.* 100, 406-419

Pellicori, S. *et al.* 1973 Pioneer imaging photopolarimeter optical system. *Applied Optics* 12, 1246-1258

Peterson, R. A., and Horn, L. H. 1977 An evaluation of 500 mb height and geostraphic wind fields derived from Nimbus 6 sounding, *Bull. Amer. Met. Soc.*, 58, 1195-1201

Phillips, N. A. 1976 The impact of synoptic observations and analysis system on flow pattern forecasts, *Bull. Amer. Met. Soc.* 57, 1225-1240

Pick, D. R. 1973 Scientific assessment of the selective chopper radiometer flown on the Nimbus V satellite. *Appl. Opt.*, 12, 303-312

Pidgeon, C. R. and Smith, S. D. 1964 *J. Opt. Soc. Am.* 54, 1459-66

Pirraglia, J. A. *et al.* 1981 Global thermal structure and dynamics of Saturn and Jupiter from Voyager infrared measurements. *Nature* 292, 675-677

Platt, C. M. R., Reynolds, D. R. and Abshire, N. L. 1980 Satellite and lidar observations of the albedo, emittance and optical depth of cirrus compared to model calculations. *Mon. Weath. Rev.* 108, 195-204

Pollack, J. B. *et al.* 1980 Distribution and sources of UV absorption in Venus atmosphere. *J. Geophys. Res.* 85, 8141-8150

Prabhakara, C., Chang, H. D. and Chang, A. T. C. 1982 Remote sensing of precipitable water over the oceans from Nimbus 7 microwave measurements. *J. Appl. Met.* 21, 59-68

Prabhakara, C., Dalu, G. and Kunde, V. G. 1974 Estimation of sea surface temperature from remote sensing in the 11 - 13 μm window region. *J. Geophys. Res.* 79, 5039-5044

Prabhakara, C. *et al.* 1976 The Nimbus-4 infrared spectroscopy experiment 3. Observations of the lower stratospheric thermal structure and total ozone. *J. Geophys. Res.* 81, 6391-6399

Prabhakara, C., Dalu, G., Lo, R. C. and Nath, N. R. 1979 Remote sensing of seasonal distribution of precipitable water-vapour over the oceans and the inference of boundary layer structure. *Mon. Weath. Rev.* 107

Purdom, J. F. W. 1982 Subjective interpretation of geostationary satellite data for now-casting, in *Nowcasting* (ed. Browning) Academic Press, 149-166

Quiroz, R. S. *et al.* 1975 A comparison of observed and simulated properties of sudden stratospheric warmings. *J. Atmos. Sci.* 32, 1723-1736

Raschke, E., Moller, F and Bandeen, W. R. 1968 Diese Arbeit erschien inMeddelanden, ser B no 28 *Sveriges Meteorologiska och Hydrologiska Institut*, Stockholm 42-57

Rasool, S. I. and Stewart, R. W. 1971 Results and interpretation of the S-band occultation experiments on Mars and Venus. *J. Atmos. Sci.* 28, 869-878

Ratcliffe, J. A. 1959 *The Magneto-ionic Theory and its Application to the Ionosphere* (London: Cambridge University Press)

Rawcliffe, R. D., Meloy, G. E., Friedman, R. M. and Rogers, E. H. 1963, Measurement of vertical distribution of ozone from a solar orbiting satellite. *J. Geophys. Res.* 68, 6425-6429

Rawcliffe, R. D. and Elliot, D. D. 1966 Latitudinal distribution of ozone at high altitudes deduced from a satellite measurement of the earth's radiance at 2840A. *J. Geophys. Res.* 71, 5077-5089

Rees, D. 1980 Global wind determination by interferometric techniques in the stroposphere and mesosphere. *Proceedings of Charm Workshop, Corsica.* European Space Agency, Paris, Publication SP-150, 193-206

Rees, D. 1982 The mission of the Dynamics Explorer satellite. *Bull. Inst. Phys.* 33, 404-406

Reigler, G. R., Drake, J. F., Liu, S. C. and Circerone, R. J. 1976 Stellar occulation measurements of atmospheric ozone and chlorine from OAO 2, *J. Geophys. Res.* 81, 4997-5001

Reigler, G. R., Atreya, S. K., Donahue, T. M., Liu, S. C. and Wasser, B. 1977 *Geophys. Res. Lett.* 4, 145-148

Reynolds, D. W. and Vonder Haar, T. M. 1976 A bispectral method for cloud parameter determination. *Mon. Weath. Rev.* 105, 446

Reynolds, D. W., McKee, T. B. and Danielson, K. S. 1978 Effect of cloud size and cloud particles on satellite observed brightness. *J. Atmos. Sci.* 35, 160-164

Reynolds, D. W. and Smith, F. A. 1979 Detailed analysis of composited digital radar and satellite data. *Bull. Am. Met. Soc.* 60, 1024-1037

Rodgers, C. D. 1970 Remote sounding of the atmospheric temperature profile in the prescent of cloud. *Quart. J. R. Met. Soc.* 96, 654-666

Rodgers, C. D. 1971 Some theoretical aspects of remote sounding in the earth's atmosphere. *J. Quant. Spectrosc. Radiat. Transfer* 11, 767-777

Rodgers, C. D. 1976a Retrieval of atmospheric temperature and compositions from remote measurements of thermal radiation. *Rev. Geophys. Space. Phys.* 14, 609-624

Rodgers, C. D. 1976b In: *Inversion Methods in Atmospheric Remote Sounding* (ed. Deepak), 117-138, New York: Academic Press

Rodgers, C. D. 1976c Evidence for the five-day wave in the upper stratosphere. *J. Atmos. Sci.* 33, 710-711

Rodgers, C. D. 1977 Statistical principles of inversion theory. In *Inversion methods in atmospheric remote sounding,* (ed. Deepak), 117-138 Academic Press.

Rodgers, C. D. 1977 Morphology of upper atmosphere temperatures. In *Dynamical and chemical coupling,* (ed. Grandal and Holtet). D. Reidel

Rodgers, E., Siddalingaiah, H., Cheng, A. T. C. and Wilheit, T. 1979 *J. Appl. Meteorol.* 18, 978-991

Rosenkranz, P. W., Komichak, M. J. & Staelin, D. H. 1982 A method for estimation of atmospheric water vapour profiles by microwave radiometry. *J. Appl. Met.* 21, 1364-1370

Rosenkranz, P. W., Staelin, D. H. and Grody, N. C. 1978 Typhoon June (1975) viewed by a scanning microwave spectrometer, *J. Geophys. Res.* 83, 1857-1968

Rossow, W. *et al.* 1980 Cloud morphology and motions from Pioneer Venus images. *J. Geophys. Res.* 85, 8107-8128

Ruff. I. *et al.* 1968 *J. Atmos. Sci.* 25, 323-32

Russell, P. J. M. and Drayson, S. R. 1972 The inference of atmospheric ozone using satellite horizon measurements in the 1042 cm^{-1} band. *J. Atmos. Sci.* 29, 376

ussell, J. M. III and Gille, J. C. 1978 In *The Nimbus 7 User's Guide* (Ed. C. R. Madrid) 71-103 Greenbelt, Md: National Aeronautics and Space Administration

utherford, I. 1976 An operational three-dimensional multi-variate statistical objective scheme. Proceedings JOC study group conference on four-dimensional data assimilation. *The GARP programme on numerical experimentation, Report No 11*, 98-121

Samuelson, R. *et al.* 1975 Venus cloud properties: infrared opacity and mass mixing ratio. *Icarus, 25*, 49163

Schlatter, T. W. and Branstator, G. W. 1978 Errors in Nimbus 6 temperature profiles and their spatial correlation. *Mon. Weath. Rev.*

Schmugge, T. 1976 Remote sensing of soil moisture. *Proceedings of the symposium on meteorological observations from space, Philadelphia,* COSPAR 1976, 11-120

Schofield, J. T., and Taylor, F. W. 1983. The mean, retrieved, solar fixed temperature and cloud top structure of the middle atmosphere of Venus. *Quart. J. R. Met. Soc.* 109, 57-80

Schofield, J. T. and Taylor, F. W. 1982*a*. Net global thermal emission from the Venusian atmosphere. *Icarus* 52, 245-262

Schofield, J. T. and Taylor, F. W. 1982*b*. The global distribution of water vapour in the middle atmosphere of Venus. *Icarus* 52, 263-278

Schwalb, A. 1972 *Modified version of the improved TIROS operational satellite (ITOS D-G)* NOAA Technical Memorandum, NESS 35 (Washington D.C.)

Scott, N. A. and Chedin, A. 1971 A least squares procedure applied to the determination of atmospheric temperature profiles from outgoing radiance, *J. Quant. Spectrosc. Radiat. Transfer* 11, 405

Sekihara, K. and Walshaw, C. D. 1969 *Ann. Geophys.* 25, 233-241

Sellers, W. D. and Yarger, D. N. 1969 The statistical prediction of the vertical ozone distribution. *J. Appl. Meteor* 8, 357

Shafrin, Yu. A. 1970 *Isv. Atm. Ocean. Phys.* 6, 696-703

Shannon, C. E. and Weaver, W. 1949 The Mathematical Theory of Communication. The University of Illinois Press

Sikdar, D. N. and Suomi, V. E. 1972 On the remote sensing of mesoscale tropical convection intensity from a geostationary satellite, *J. Appl. Met.* 11, 37-43

Simmons, A. J. 1974 Planetary-scale disturbances in the polar winter stratosphere. *Quart. J. R. Met. Soc.* 100, 76-108

Singer, S. F. and Wentworth 1957 A method of determination of the vertical ozone distribution from a satellite, *J. Geophys. Res.* 62, 229-309

Sissala, J. F. Ed., 1975 *The Nimbus 6 User's Guide.* NASA (Goddard Space Flight Center)

Slysh, V. I. 1976 Identification of the CO molecule in the radiation spectrum of the Venusian night-time sky luminescence of Venus from Venera 9 and 10. *Kosicheskie Issledovaniya* 14, 796-798

Smith, S. D. 1961 *Quart. J. R. Met. Soc.* 87, 431-434

Smith, S. D., Collis, M. J. and Peckham, G. 1972 The measurement of surface pressure from a satellite. *Quart. J. R. Met. Soc.* 98, 431-433

Smith, S. D. and Pidgeon 1964 *Mem. Soc. R. Sci. Liege* 9, 336-49

Smith, W. L. 1967 An iterative method for deducing tropospheric temperature and moisture profiles from satellite radiation measurements. *Mon. Weath. Rev.* 95, 363

Smith, W. L. 1968 An improved method for calculating tropospheric temperature and moisture from satellite radiometer measurements. *Mon. Weath, Rev.* 96, 387-396

Smith, W. L. 1970 Iterative solution of the radiation transfer equation for the temperature and absorbing gas profile of an atmosphere. *Appl. Opt,* 9, 1993

Smith, W. L., Woolf, H. M. and Jacob, W. J. 1970 A regression method for obtaining real time temperature and geopotential height profiles from satellite spectrometer measurements and its application to Nimbus 3 SIRS observations. *Mon. Weath. Rev.* **98**, 582-603

Smith, W. L. *et al.* 1972 The Infrared Temperature Profile Radiometer (ITPR) experiment. *Nimbus-5 User's Guide*, 107-130 NASA (Goddard Space Flight Center)

Smith, W. L., Woolf, H. M. and Fleming, H. E. 1972 Retrieval of atmospheric temperature profiles from satellite measurements for dynamical forecasting. *J. Appl. Meteor.* **11**, 113

Smith, W. L., Staelin, D. H. and Houghton, J. T. 1973 Intercomparison and amalgamation of Nimbus 5 and infrared and microwave temperature profile data. *Proceedings of international symposium on meteorological satellite, C.N.E.S. Paris*, May 1973, 139-146

Smith, W. L., Staelin, D. H. and Houghton, J. T. 1974 Vertical temperature profiles from satellites: results from second generation instruments aboard Nimbus 5. In *Approaches to earth survey problems through space techniques* (ed. Bock), 123-143. Academic-Verlag, Berlin.

Smith, W. L. *et al.* 1975 The high resolution infrared radiation sounder (HIRS) Experiment. *Nimbus-6 User's Guide*, 37-58 NASA (Goddard Space Flight Centre)

Smith, W. L. and Woolf, H. M. 1976 The use of eigenvectors of statistical covariance matrices for interpreting satellite sounding radiometer observations, *J. Atmos. Sci.* **33**, 1127-1140

Smith, W. L. *et al.* 1977 Nimbus 6 earth radiation budget experiment. *Appl. Optics* **16**, 306-318

Smith, W. L. *et al.* 1978 *Satellite sounding applications to mesoscale meteorology*, Paper presented at COSPAR Symposium, Innsbruck, 1978

Smith, W. L. and Platt, C. M. R. 1978 Comparison of satellite deduced cloud heights with indications from radiosonde and ground based laser measurements. *J. Appl. Met.* **17**, 1976

Smith, W. L., Hayden, C. M., Woolf, H. M., Howell, M. B. and Nagle, F. W. 1979 Satellite sounding applications to mesoscale meteorology. In 'Remote Sounding of the Atmosphere from Space', (ed. H.-J. Bolle) *Pergamon Press*

Smith, W. L. *et al.* 1981 First Sounding Results from VAS-D. *Bull. Amer. Met. Soc.* **62**, 232-236

Smith, W. L., Suomi, V. E., Zhou, F. X. and Menzel, W. P. 1982 Nowcasting applications of geostationary satellite atmospheric sounding data. In *Nowcasting* (ed. K. H. Browning). *Academic Press* 123-135

Staelin, D. H., Barath, F. T., Blinn, J. C. and Johnson, E. J. 1972 The Nimbus E microwave spectrometer (NEMS) experiment. *Nimbus-5 User's Guide*, 141-157. NASA (Goddard Space Flight Centre)

Staelin, D. H. *et al.* 1975 The scanning microwave spectrometer (SCAMS) experiment, *Nimbus-6 User's Guide*, 59-86. NASA (Goddard Space Flight Centre).

Staelin, D. H. *et al.* 1976 Remote sensing of atmospheric water-vapour and liquid water with the Nimbus 5 microwave spectrometer. *J. Appl. Met.* **15**, 1204-1214

Staelin, D. H. *et al.* 1977 Microwave spectroscopic imagery of the earth. *Science* **197**, 991-993

Steranka, J., Allison, L. J. and Salomonson, V. V. 1973 Application of Nimbus 4 THIR 6.7 μm observations to regional and global moisture and wind field analyses, *J. Appl. Met.* **12**, 386-395

Stephens, G. L., Cambell, G. G. and Vonder Haar, T. H. 1981 Earth radiation budgets. *J. Geophys. Res.* **88**, 9739-9760 uchman, D. and Martin, D. W. 1976 Wind sets from SMS images: an assessment of quality for GATE, *J. Appl. Met.* **15**, 1265-1278

uchman, D., Martin, D. W. and Sikdar, D. N. 1977 Deep convective mass transports: an estimate from a geostationary satellite, *Mon. Weath. Rev.*, **105**, 943-955

Suomi, V. E. 1969 In *The Global Circulation of the Atmosphere* (ed. G. Corby) 222-234. London: Royal Meteorological Society

Suomi, V. E. 1975 Cloud motions on Venus. In *The atmosphere of Venus*, (ed. J. E. Hansen), 42-58. NASA SP-382

Surmont, J. and Chen, Y. M. 1973 Numerical solution of a nonlinear radiation transfer equation with inadequate data. *J. Comput. Phys.* **13**, 288

Swindell, W. and Doose, L. R. 1974 The imaging experiment on Pioneer 10. *J. Geophys. Res.* **79**, 3634-3644

Tapley, B. D. *et al.* 1979 Seasat altimeter calibration–initial results. *Science* **204**, 1410-1412

Taylor, F. W. 1972 Temperature sounding experiments for the Jovian planets. *J. Atmos. Sci.* **29**, 950-958

Taylor, F. W. *et al.* 1972 Radiometer for remote sounding of the upper atmosphere. *Applied Optics.* **11**, 135-141

Taylor, F. W. 1974 Remote temperature sounding in the presence of cloud by zenith scanning. *Appl. Opt.* **13**, 1559-1566

Taylor, F. W. 1975 Interpretation of Mariner 10 infrared observations of Venus. *J. Atmos. Sci.* **32**, 1101-1106

Taylor, F. W. *et al.* 1979*a* Polar clearing in the Venus clouds observed from Pioneer Venus orbiter. *Nature* **279**, 5714-5716

Taylor, F. W. *et al.* 1979*b* Infrared radiometer for the Pioneer Venus Orbiter: Instrument Description. *Appl. Optics*, **18**, 3893-3900

Taylor, F. W. *et al.* 1980 Structure and meteorology of the middle atmosphere of Venus: Infrared remote sensing from the Pioneer Orbiter. *J. Geophys. Res.* **85**, 7963-8006

Taylor, F. W. *et al.* 1981 Comparative aspects of Venusian and Terrestrial meteorology. *Weather*, **36**, 34-40

Taylor, F. W. *et al.* 1982 Pioneer Venus atmospheric observations. *Phil. Trans. Roy. Soc. Lond.*, in press

Taylor, F. W. 1983 The Pressure Modulator Radiometer. Spectrometric Techniques vol. 4. (ed. G. Vanasse) *Academic Press*, New York

Thrush, B. A. 1979 Aspects of the chemistry of ozone depletion. *Phil. Trans. Roy. Soc.* A**290**, 505-514

Tikhonov, A. N. 1963 On the solution of incorrectly stated problems and a method of regularization. *Dokl. Acad. Nauk. USSR* **151**, 501

Tomasko, M. T. *et al.* 1978 Photometry and polarimetry of Jupiter at large phase angles. *Icarus* **33**, 558-592

Tomasko, M. T. *et al.* 1980 The thermal balance of Venus in light of the Pioneer Venus mission. *J. Geophys. Res.* **85**, 8187-8199

Toth, R. A. and Farmer, C. B. 1971 *Joint Conf. Sensing of Environmental Pollutants*. American Institute of Aeronautics and Astronautics AIAA Paper no. 71-1109

Tracton, M. S. and McPherson, R. D. 1977 On the impact of radiometric sounding data upon operational numerical prediction at MNC. *Bull, Amer. Met. Soc.* **58**, 1201-1209

Turchin, V. F. and Nozik, V. Z. 1969 Statistical regularization of the solution of incorrectly posed problems. *Izv. Acad. Sci. USSR, Atmos. Oceanic Phys.* **5**, 14

Twomey, S., 1961 On the deduction of the vertical distribution of ozone by ultraviolet spectral measurements from a satellite. *J. Geophys. Res.* **66**, 2153-2162

Twomey, S., 1963 On the numerical solution of Fredholm integral equations of the first by the inversion of the linear system produced by quadrature. *J. Ass. Comput. Mach.* 10, 97

van Loon, H. *et al.* 1972 Half-yearly wave in the stratosphere. *J. Geophys. Res.* 77, 3846-3855

Venkateswaran, S. V., Moore, J. G. and Krueger, A. J. 1961 Determination of the vertical distribution of ozone by satellite photometry, *J. Geophys. Res.* 66, 1751-1771

Vetlov, I. P. 1979 Applications of satellites in hydrometeorology. In *'Use of data from meteorological satellites'.* Publ. SP.-143 European Space Agency, Paris, 15-28

Vonder Haar, T. H. and Suomi, V. E. 1971 *J. Atmos. Sci.* 28, 305-14

Wark, D. Q., 1960 On indirect temperature soundings of the stratosphere from satellites. *J. Geophys. Res.* 66, 77

Wark, D. Q., Yamamoto, G. and Lienesch, J. H. 1962 *J. Atmos. Sci.* 19, 369-84

Wark, D. Q., and Hilleary, D. T. 1969 Atmospheric temperature: successful test of remote probing. *Science* 165, 1256-1258

Waters, J. W. *et al.* 1975 The shuttle imaging microwave system experiment. *Proceedings of IEECC National Telecommunications Conference, New Orleans,* 1975

Webster, W. J. Jr., Wilheit, T. T., Chang, T. C., Gloerse, P. and Schmugge, T. J. 1975 A radio picture of the earth, *Sky and telescope* 49, 14-16

Webster, P. J. *et al.* 1977 Equatorial waves in the upper stratosphere. *Proc. IAMAP/IAGA Assembly,* Seattle

Westwater, E. R. and Strand, O. N. 1968 Statistical information content of radiation measurements used in indirect sensing. *J. Atmos. Sci.* 25, 750-8

Wilheit, T. T., Theon, J. S., Shenk, W. E., Allison, W. J. and Rodgers, E. B. 1976 Meteorological interpretations of the images from the Nimbus 5 electrically scanned microwave radiometer. *J. Appl. Met.* 15, 166-172

Wilheit, T. T., Change, A. T., Rao, M. S., Rodgers, E. B. and Theon, J. S. 1977 A satellite technique for mapping rainfall rates over the oceans. *J. Appl. Met.* 16, 551-560

Williams, F. L. and McCandless, W. W. Jr., 1976 Seasat, an ocean observation satellite. *Proceedings of the symposium on meteorological observations from space,* 361-367 Philadelphia, COSPAR

Willson, R. C. 1980 Active cavity radiometer type V. *Applied Optics.* 19, 3256-3257

Willson, R. C. and Hudson, H. A. 1981 Variations of solar irradiance. *Astrophysics J,* 244, L185-L189

Woerner, C. V. and Cooper, J. E. 1977 *Earth radiation budget satellite system studies,* NASA Technical Memorandum TM X-72776

Woiceshyn, P. 1974 Global seasonal atmospheric fluctuations on Mars. *Icarus* 22, 324

World Meteorological Organization 1969 The planning of the first GARP Global Experiment, GARP Publication Series No. 3. WMO, Geneva

Yamamoto, G. 1961 Numerical method for estimating the stratospheric temperature distribution from satellite measurements in the CO_2 band. *J. Met.* 18, 581

INDEX AND GLOSSARY